STUDENT UNIT GUIDE

NEW EDITION

WJEC A2 Geography Unit G3

Contemporary Themes and Research in Geography

Nicky King

Series editor: David Burtenshaw

PHILIP ALLAN

With thanks to Gill Miller of Chester University for her contribution to Themes 6(a) and 6(b)

Philip Allan Updates, an imprint of Hodder Education, an Hachette UK company, Market Place, Deddington, Oxfordshire OX15 0SE

Orders
Bookpoint Ltd, 130 Milton Park, Abingdon, Oxfordshire OX14 4SB
tel: 01235 827827
fax: 01235 400401
e-mail: education@bookpoint.co.uk
Lines are open 9.00 a.m.–5.00 p.m., Monday to Saturday, with a 24-hour message answering service. You can also order through the Philip Allan Updates website: www.philipallan.co.uk

ISBN 978-1-4441-6203-5

First printed 2012
Impression number 5 4 3 2 1
Year 2016 2015 2014 2013 2012

Cover photo: Aania/Fotolia

Printed in Dubai

Hachette UK's policy is to use papers that are natural, renewable and recyclable products and made from wood grown in sustainable forests. The logging and manufacturing processes are expected to conform to the environmental regulations of the country of origin.

This material has been endorsed by WJEC and offers high quality support for the delivery of WJEC qualifications. While this material has been through a WJEC quality assurance process, all responsibility for the content remains with the publisher.

P2031

Contents

Content Guidance

Questions & Answers

Getting the most from this book

Questions & Answers

Exam-style questions

Examiner comments
on the questions
Tips on what you
need to do to gain full
marks, indicated by the
icon ℮.

Sample student
answers
Practise the questions,
then look at the student
answers that follow
each set of questions.

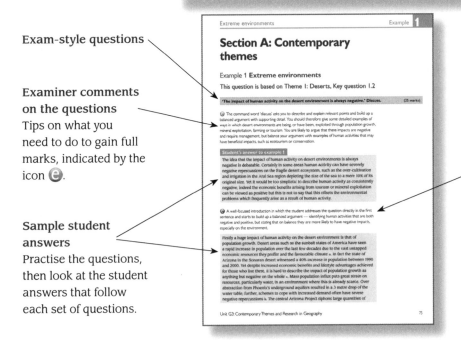

Examiner commentary
on sample student
answers
Find out how many marks
each answer would be
awarded in the exam and
then read the examiner
comments (preceded by
the icon ℮) following
each student answer.
Annotations that link
back to points made in
the student answers show
exactly how and where
marks are gained or lost.

About this book

The purpose of this guide is to help you understand what is required to do well in Unit G3: Contemporary Themes and Research in Geography. The specification is available on the WJEC website: **www.wjec.co.uk**.

This guide is divided into two sections. The Content Guidance section sets out the *bare essentials* of the specification for this unit. Diagrams are used to help your understanding; many are simple to draw and could be used in the exam. The Questions & Answers section provides guidance on how to approach the unit test and includes examples of the types of question that you will see in the examination. Sample answers of differing standards are provided, as well as examiner's comments on how to tackle each question and on where marks have been gained or lost in the sample answers.

Content Guidance

Section A: Contemporary themes

Theme 1 Extreme environments

1.1 What are the characteristics of a desert environment that make it extreme?

The climatic, biotic and soil characteristics of a desert environment

Hot deserts are located in the subtropical zone between 15° and 35° north and south of the Equator. Examples include the Sahara in north Africa and the Kalahari Desert in the south. The hot desert climate is characterised by high **insolation** rates, subsiding air and high-pressure conditions with outblowing winds. These factors result in exceptionally dry conditions, generally clear skies, and very high evaporation rates, with temperatures during the day averaging over 40°C. On the western land margins, cold ocean currents deprive inblowing marine winds of their moisture before they reach land, contributing to the exceptionally dry conditions and leading to fog conditions offshore, for example, off northern Chile and Namibia. Occasional incursions of the **Inter-Tropical Convergence Zone (ITCZ)** into desert latitudes may allow the penetration of moist winds bringing very heavy, but short-lived, convectional downpours of rain. Average precipitation is 250 mm, but effective rainfall is less, primarily due to the very high evaporation rates. **Diurnal** contrasts are more significant than **seasonal** variations as rapid radiation at night, due to clear skies, can result in below-freezing temperatures and heavy morning dew in contrast to the high temperatures and very low humidity conditions during the day.

The productivity of **grey desert soils** is extremely low because the lack of water and high evaporation rates lead to salts accumulating and crystallising on the soil surface. The **organic** layer is very thin as **primary productivity** to produce **litter** is extremely low.

The links between climate, biotic and soil characteristics

Special adaptations are required by biota to overcome the extreme climatic and soil characteristics. The combination of extremes of temperature, potential water loss exceeding annual precipitation and thin, saline soils create challenges for organisms. Adaptation to these extremes includes surviving as an **ephemeral** that takes advantage of short-lived convectional downpours by germinating, growing, flowering and seeding within 20–30 days. Most desert organisms are adapted to withstand or avoid water stress. **Succulents**, such as the cacti of the Americas, have thick cuticles, a low surface-area-to-volume ratio and sunken stomata which open at night to minimise transpiration. These adaptations enable them to survive above ground throughout the year.

Animals have to adapt to the scarcity of water and food and the temperature extremes. Camels have thick fur to insulate their bodies from the sun, splayed hooves for walking on mobile sands and humps that store fat that provides both energy and water on being respired.

1.2 How is human activity causing pressure on the desert environment?

The threats that are posed by population growth, mineral exploitation, farming and tourism. The positive and negative outcomes of human activity

The desert biome is under threat from a variety of human pressures including population growth, mineral exploitation, farming and tourism. However, the impacts of such activities can be both positive and negative.

Population growth is occurring in desert environments including the Sahara Desert at an average rate of 1.6% per annum. Population growth and greater demand on the land in Tunisia have led to a decline in soil fertility, the lowering of the water table, the silting of dams, and increased flood risk due to new buildings and infrastructure.

The Namib Desert has been sensitive to human activities for decades. The impacts of prospecting and **mineral exploitation** for lithium, beryllium, vanadium and tantalum, as well as more common minerals such as tin, zinc, lead and diamonds have left permanent scars such as the ghost town of Kolmanskop.

In the Sonoran Desert, irrigated **farming** and ranching have been ecologically and economically unsustainable. Vegetation clearing, followed by later field abandonment, the build-up of saline and alkaline soil crusts, pesticide and herbicide use, nitrate fertiliser contamination of stream flows and aquifers, as well as nitrogen enrichment of adjacent wildlands, the introduction and spread of exotic weeds, plant diseases and insects, have all impacted negatively on the desert environment.

On the other hand, agriculture can sometimes have a positive effect on wildlife and fieldside wild plant populations when agriculture is practised on a modest scale, without pesticide and herbicide use.

The increase in **tourism** in the United Arab Emirates, especially involving uncontrolled activity by off-road recreational vehicles (ORVs), and quad and motocross bikes, has had negative impacts on the desert environment and its wildlife. In contrast, the Sinai Peninsula in Egypt offers ecotourists the opportunity to trek with the Bedouin and their camels, causing minimal impact.

Examiner tip
Note that it is the inter-relationships between the desert climate, soils and vegetation characteristics that illustrate the extremity of desert environments.

Examiner tip
Make the link between human activity and the effect it has on the fragile and special qualities of the desert environment clear and detailed. Use located examples to illustrate your point.

Examiner tip
Examiners are looking for an ecological emphasis in your answers rather than a focus on wider socio-economic impacts such as employment opportunities. The impacts of human activity you refer to should be confined to desert environments.

1.3 What are the strategies that can be used to manage human activity in deserts?

Strategies that attempt to conserve the desert environment, alleviate the impacts of human activity, control the use of the desert environment and monitor the impacts of human activity

In Tunisia, where 21% of the country is classified as desert and 17.2% is at risk from desertification, a range of strategies have been implemented to manage human activity in the desert environment. Soil and water conservation techniques such as terracing, the construction of earth dams, and slope protection are being implemented by agronomists and hydraulic engineers. Soils degraded through human activity are being treated chemically and enriched with organic matter to counteract their alkalinity. Energy consumption is being altered and rationalised in rural areas to reduce the use of brushwood. In order to monitor the impact of human activity, land degradation maps are being compiled using remote sensing and with the development of a national observatory of land degradation.

Knowledge check 1

What strategies can be used to prevent soil erosion in deserts?

The **Sonoran Desert Conservation Plan** is encouraging ranching as an extensive but low-intensity form of land use. Conserving ranches that are currently in operation provides the best way to preserve the integrity of vast tracts of connected, unfragmented open space and wildlife habitat.

The role of local, national and international groups in the management of the desert environment

In Tunisia, various groups are working at different levels to manage the desert environment and avoid it spreading. At the international level, UN institutions such as the FAO (Food and Agriculture Organization) and UNCCD (United Nations Convention to Combat Desertification) have proposed plans that have been adopted at the national level. Tunisia has created a national action plan involving international, national and local organisations. The groups involved at the national level include the Agriculture Ministry, the Ministry for Economic Development and the Environment, and research institutes such as the National Agency for the Protection of the Environment and the National Commission for Sustainable Development. The plan also involves experts and farmers and provides a skills and education action framework. Each region has a development office responsible for environmental protection. The Tunisian national action plan is proving to be effective, but Tunisia has more resources at its disposal than most other African countries.

1.4 What are the characteristics of the arctic and alpine tundra environment that make it extreme?

The climatic, biotic and soil characteristics of a tundra environment

The high latitude of **arctic tundra environments** results in very low temperatures with average temperatures ranging between -5°C and -10°C, long dark winter months when temperatures fall below -20°C and high winds. Tundra regions are dominated by high pressure and subsiding air leading to low mean annual precipitation of below 150 mm. During the short growing season daylight hours are long, but the sun's angle is so low that temperatures rarely rise above 10°C. In summer the ground is permanently frozen apart from the top 50 cm.

The links between climate, biotic and soil characteristics

In arctic areas the growing season is so short that few plants are annuals. Vegetation is dominated by mosses, lichens, grasses and dwarf shrubs. Most plants have short roots so as to avoid the permafrost and small leaves so as to limit transpiration. Dwarf willow and stunted birch trees, with their crowns distorted by the wind, grow adjacent to seasonal rivers but only to a maximum height of 30 cm. Plant growth and flowering can be spectacular and colourful during the short growing season. Large numbers of insects appear, giving a few weeks of high productivity. Limited plant growth results in a small amount of litter and the lack of soil biota results in the slow decomposition of organic matter to give a thin layer of peat. When meltwater percolates downwards in the late spring, humic acid releases iron. Underlying permafrost acts as an impermeable layer, causing waterlogging and **gleying**. Bedrock weathered by freeze–thaw action is raised to the surface by frost heave, preventing the formation of soil horizons.

Alpine tundra environments experience low temperatures, high **orographic** precipitation (usually in the form of snow) and high wind speeds. In lower latitudes, alpine tundra environments have a higher diurnal temperature range and frequent freeze–thaw cycles. Soils are usually thin with large amounts of loose rock. Small, brightly coloured alpine plants, such as gentians, grow slowly. Biotic diversity attracts tourists to the Alps at flowering time.

1.5 How is human activity causing pressures on the arctic and alpine tundra environment?

The threats that are posed by mineral exploitation, airborne pollution, global warming and tourism. The positive and negative outcomes of human activity

The tundra biome is under threat from a variety of human pressures including mineral exploitation, airborne pollution, global warming and tourism.

Tundra A biome where tree growth is hindered by low temperatures and short growing seasons.

Examiner tip
Provide climatic data for a tundra location to exemplify its extreme climate characteristics.

Knowledge check 2

What adaptations do plants and animals have to enable them to live in tundra environments?

Around Prudhoe Bay in Alaska, the heat produced by buildings and pipelines associated with oil extraction causes permafrost melting. This results in frost heave in the active layer and damage to plants, which take many years to become re-established.

The fragile ecosystems of periglacial environments mean they have a low tourist-carrying capacity. Kuujjuaq in Quebec province, Canada, attracts a small number of tourists interested in hunting and fishing. Nevertheless, irreversible damage has been caused by tourists walking through the tundra.

Global atmospheric circulation patterns are such that the air in industrial areas in northern mid-latitudes becomes polluted and transported to tundra regions, where impurities are deposited.

Tourism in the Alps has grown significantly with the result that tourist and other commercial vehicles are causing dangerous air pollution levels in many alpine valleys. Large areas of permafrost are experiencing greater surface melting associated with the increased atmospheric CO_2 produced by human activity. This creates further positive feedback as bare ground has a lower albedo, which raises surface temperatures and leads to further melting. This melting releases large quantities of methane and leads to further global warming.

1.6 What are the strategies that can be used to manage human activity in arctic and alpine tundra environments?

Strategies that attempt to conserve the tundra environment, alleviate the impacts of human activity, control the use of the tundra environment and monitor the impacts of human activity

The Canadian government has given landmark status to Tuktoyaktuk, an Inuit town in the Northwest Territories that has 1,400 **pingos**, to protect these landforms and the wider area from tourism.

The role of local, national and international groups in the management of the tundra environment

Various groups are working at different levels to manage the tundra environment. At the international level there are global warming initiatives, e.g. Kyoto (1997) and Copenhagen (2009). The Wildlife Management Advisory Council (WMAC) works with the governments of the Canadian province of Yukon and the US state of Alaska, and the Canadian national government, on conserving the wildlife, habitat and traditional use of the Yukon North Slope.

Examiner tip
You will be rewarded for providing details of strategies to manage human activity in tundra environments that are implemented at a local, national and international level.

Summary

After studying this topic, you should be able to:

- describe and explain the climatic, biotic and soil characteristics of a desert environment and understand how these are interrelated

- provide detailed and located examples of human activity and explain how these are placing pressure on the fragile and special qualities of the desert environment

- give examples of strategies used by local, national and international groups to manage human activity in deserts

- describe and explain the climatic, biotic and soil characteristics of the tundra environment and understand how these are interrelated

- provide detailed and located examples of human activity and explain how these are placing pressure on the fragile and special qualities of the tundra environment

- give examples of strategies used by local, national and international groups to manage human activity in tundra environments

Theme 2(a) Glacial landforms and their management

1.1 What is a glacial system and what are the dynamics of glacial environments?

Glacier mass balance

The inputs to and outputs from a glacier are not constant, but vary over both short and long timescales. The glacier system constantly adjusts to changes in the balance between **accumulation** and ablation and this is reflected in the **mass balance** of a glacier. If accumulation exceeds ablation a glacier gains mass (**positive mass balance**). If there is more ablation than accumulation a glacier has a **negative mass balance**.

The impact of climate change on glacier budgets

Glaciers have shown periods of expansion and retreat as climate changes have shifted the net balance to either positive (colder conditions) or negative (warmer conditions).

The relationship between climate fluctuations and the geomorphological work done by ice

The geomorphological work done by ice and the glacial landforms created can be linked to global events that changed the climate. Glacial advance occurs on different timescales associated with fluctuations in climate, such as during the glacials of the Pleistocene epoch and 'Little Ice Age'. This results in higher energy levels and increased erosion in the form of abrasion and plucking, creating major landforms of glacial erosion. Glacier retreat associated with inter-glacials of the Pleistocene epoch and present-day global warming result in reduced energy levels and increased glacial deposition, fluvio-glacial processes and associated landforms.

Ablation The collective loss of water from a glacier by processes such as surface melting, evaporation and wind-blown snow.

Examiner tip

Support your explanations with examples. An example of climatic fluctuations influencing glacial processes and landforms is the 'Younger Dryas', a short-lived temperature fluctuation at the end of the last glacial cycle. An influx of cold fresh water into the North Atlantic resulted in re-growth of glaciers, producing cirque moraines, e.g. Cwm Idwal.

Cold- and warm-based glaciers, their types and rates of movement

Glaciers can be classified as **cold-based** or **warm-based** depending on whether they are frozen to the underlying bedrock or not. Outside of the polar regions most glaciers are warm-based. However, large glaciers can be cold-based in their upper regions and warm-based near their margins when they extend into temperate climatic zones. Slow rates of **accumulation** and **ablation** associated with glaciers in cold continental climates result in a smaller imbalance between accumulation and ablation and slower ice movement. Glaciers in temperate–maritime climates have greater snowfall in winter and experience more rapid ablation in summer; consequently, the glacier moves more rapidly. There is much more erosion under warm-based glaciers than under cold-based glaciers.

1.2 What are the processes of glacial weathering and erosion and what are the resultant landforms?

Weathering and erosion in the glacial zone

Weathering is the disintegration and decomposition of rocks in situ. The relatively high humidity combined with relatively low temperatures that fluctuate above and below freezing make **freeze–thaw** weathering, a **physical weathering** process, predominant. The low temperatures make **chemical weathering** less important. Generally freeze–thaw weathering contributes to the formation of angular features in glacial environments, such as **arêtes**. Freeze–thaw weathering on slopes produces rock slides leading to **screes**, but higher temperatures in summer may lead to mudflows and soil creep, known as **solifluction**. Freeze–thaw weathering and mass movement of scree contribute directly to the formation of various types of **moraines** and provide glaciers with the abrasive material that makes these agents effective. Weathering is very important as it allows the ice to move material that has already been loosened.

The processes of **glacial erosion** can be divided into three categories: glacial **plucking**, glacial **abrasion** and glacial **meltwater** erosion.

- **Plucking** occurs when ice freezes around protruding rocks that are then plucked away as the ice moves. This process is particularly effective on well jointed rocks and in previously weathered areas, such as the backwalls of cirques.
- **Abrasion** involves the 'sandpapering' effect of angular material embedded in a glacier's sides and base. This is only effective when there is a continuous supply of moraine.
- Glacial **meltwater** erosion is not, strictly speaking, a glacial process, but removes material during spring and summer melting.

Landforms of glacial erosion to include macro-scale, meso-scale and micro-scale landforms

The characteristic macro-scale landforms produced by erosional processes include **cirques**, **arêtes**, **pyramidal peaks**, **glacial troughs**, **hanging valleys**, **truncated spurs** and **crag-and-tail** landforms.

Meso-scale landforms include **roches moutonnées** and **subglacial meltwater channels**.

Micro-scale landforms include **striations**.

Cirques are basin-shaped mountain hollows formed by small cirque glaciers that are characterised by rotational movement. Their steep head- and sidewalls result from freeze–thaw weathering and plucking, and their smooth and gentle floors are the result of abrasion. Under pressure from the volume of ice, cirque glaciers rotate sufficiently to move uphill and over the **rock lip** at the exit from the cirque, e.g. as at Cwm Cau on Cadair Idris mountain (Figure 1).

Knowledge check 3

Where would the erosional process of plucking be concentrated during the formation of a cirque? How does plucking influence the operation of another erosional process that is important in cirque formation?

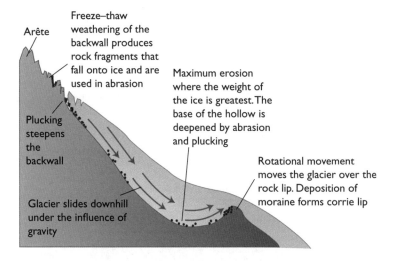

Figure 1 Cross-section to show glacial processes responsible for cirque formation

Arêtes are narrow, knife-edged ridges resulting from the headward erosion of neighbouring cirques through freeze–thaw weathering and plucking, e.g. Crib Goch, Snowdonia.

Pyramidal peaks form as a result of headward erosion by three or more cirque glaciers, e.g. Mount Snowdon.

Glacial troughs start with a pre-glacial V-shaped river valley which is deepened and has its sides steepened by the processes of freeze–thaw weathering, abrasion and plucking, to eventually form a U-shaped glacial valley, e.g. Nant Ffrancon valley, Snowdonia.

Hanging valleys relate to pre-glacial tributary streams feeding the main river channel. Glacial overdeepening of this main valley leaves the less eroded tributary valley hanging over the main trough, with its river as a waterfall cascading over the edge.

Truncated spurs indicate how powerful glacial erosion is compared to fluvial erosion, as the interlocking spurs of pre-glacial river valleys are eroded by powerful valley glaciers that follow an essentially straight course.

Crag-and-tail landforms (Figure 2) develop where a glacier overrides a mass of hard rock (the crag), which protects softer rocks in its lee. The latter form a tapered, gently sloping ridge (the tail). This is a good example of *differential erosion*.

Differential erosion
The selective erosion of landforms so that hard, resistant, unjointed bedrock is eroded more slowly than softer rocks or those with lines of weakness, such as joints and faults.

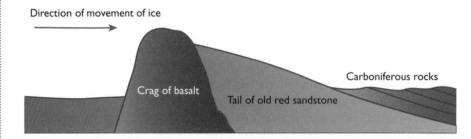

Figure 2 Crag-and-tail landforms

Roches moutonnées develop when a large block of resistant rock is eroded. The upglacier (**stoss**) slope shows evidence of abrasion in the form of striations, and slopes gently. The downglacier (**lee**) face is steeper and rougher due to plucking (Figure 3).

Subglacial meltwater channels develop where meltwater is channelled beneath a glacier, widening and deepening existing grooves, e.g. Cheriton Valley, Gower.

Striations are abrasion marks on rock surfaces that indicate the direction of ice flow.

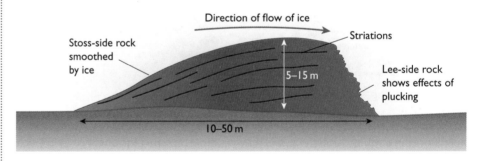

Figure 3 A roche moutonnée (section)

1.3 What are the processes of glacial transport and deposition and what are the resultant landforms?

Transport and deposition in the glacial zone

The transportation of load by glaciers is dependent on the depth of the ice, the amount of load carried and the temperature and pressure conditions that affect the 'fluidity' of the ice slope gradient. Transported material (debris) can be classified as either **supraglacial** (on the glacier surface), **englacial** (within the glacier), or **subglacial** (beneath the glacier). Subglacial debris is the most altered during transport. Deposition is dependent on changes in levels of energy in relation to the load carried. This may be affected by changes in atmospheric processes, so in periods of higher temperature and less snowfall ice movement is reduced, leading to lower energy levels and more deposition. The processes by which glaciers deposit material are complex.

Landforms of glacial and fluvioglacial deposition

The characteristic landforms produced by depositional processes include subglacially formed moraines such as **drumlins** and **till plain**, and ice-marginal moraines such as **terminal**, **recessional**, **lateral** and **medial** moraines. In periods of higher temperature when ice and snow melt occurs, fluvioglacial deposition is important and creates **eskers**, **kames** and **outwash gravels**. **Kettle holes** or **kettle lakes** may also occur when the ice beneath melts.

Drumlins are smooth, oval hills that occur in swarms, creating 'basket-of-eggs' topography'. They are likely to have formed under deep, mobile ice and their degree of elongation is related to the speed of ice movement (Figure 4). In Ireland a belt consisting of tens of thousands of tightly-packed drumlins extends from County Down in Northern Ireland, through County Mayo to Donegal Bay in the Republic of Ireland.

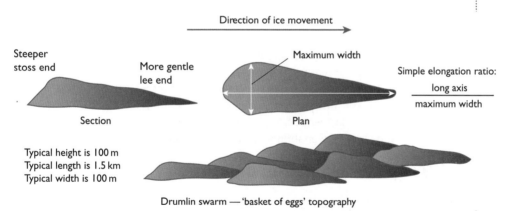

Drumlin swarm — 'basket of eggs' topography

Figure 4 Drumlins, showing stoss and lee ends and drumlins' orientation

Till plain is a combination of subglacial, englacial and supraglacial moraine released by a stagnant glacier as it melts, which leaves the debris unsorted.

Terminal moraines mark the maximum advance of the glacier or ice sheet, e.g. the Glais moraine in the Swansea Valley.

Recessional moraines form where the ice front remained stationary for a period before retreating further.

Lateral moraines are embankments of freeze–thaw weathered debris running along the sides of a glacier's valley. **Medial moraines** result from the merging of lateral moraines from two tributary glaciers joining together.

Examiner tip
Ensure that you have your own examples of all these landforms as your answers will be given credit if they contain examples that are original.

15

Eskers are winding ridges of silt, sand and gravel laid down by meltwater in a subglacial tunnel orientated approximately at right angles to the ice front.

Kames are irregular mounds of stratified sands and gravels, formed from a decaying glacier or ice sheet.

Outwash gravels are horizontally bedded sheets of gravel deposited by summer melt or deglaciation downward of the ice front.

Kettle holes are circular depressions, initially filled by meltwater, resulting from the gradual decay of a block of ice buried by overlying sediments.

1.4 What are the effects of deglaciation on the landscape?

The effects of deglaciation on the landscape to include periglacial, fluvial and sub-aerial processes

Deglaciation is the reduction in size of glaciers and ice sheets resulting from a negative mass balance. It leads to the exposure of a previously ice-covered surface. It includes the retreat of periglacial processes and landforms to higher altitudes and latitudes. In areas of low relief, important periglacial processes are **frost heaving** and **thrusting**. Associated periglacial landforms are **pingos** and **patterned ground**. On slopes, important periglacial process are **freeze–thaw weathering** and **solifluction** and associated periglacial landforms are **blockfields**, **scree slopes** and **solifluction lobes** and **benches**. Relevant geomorphological processes include **mass movement** processes (modifying valley profiles largely created by glacial erosion), fluvial processes (resulting in the infilling at the head of ribbon lakes), or weathering processes (breaking down glacial and fluvioglacial deposits). Since the last glaciation the change to temperate conditions, together with changes in **base level** due to isostatic adjustment, have significantly modified glacial landforms. Other examples of the influence of isostatic adjustment are raised erosion surfaces in glaciated uplands and rejuvenation features within glaciated valleys.

1.5 Why are glacial environments important?

The impact of glacial processes and landforms on human activity

Glacial processes impact on human activity because of the high incidence of **avalanches**, **rock falls** and other forms of mass movement such as **landslides** and **glacial outburst floods**. On 23 February 1999, 31 people were killed by an avalanche in Galtür, Austria, which took only 50 seconds to reach the village.

Glacial landforms (in areas that are currently experiencing glaciation and in formerly glaciated areas) present constraints and provide opportunities for human activity in terms of tourism, water supplies and energy, agriculture, mining and quarrying, settlement and corridors for transport.

The impact of human activities on glacial environments

Some of the impacts of human activities on glacial environments are:

- Leisure activities — winter-sports activities, including associated infrastructure such as buildings, ski lifts and road access.
- Logging activities leading to the removal of vegetation cover, which accelerates weathering and mass movement processes.
- Damming of glacial lakes for use as reservoirs for hydroelectric power schemes.
- Pollution and permafrost degradation through settlement and heat and waste disposal.
- Anthropogenic climate change, leading to the net ablation of glaciers worldwide.

Anthropogenic
Something that has been made or caused by humans.

Opportunities and limitations for human activity presented by the shift of the permafrost limit

Opportunities for human activity presented by the shift of the permafrost limit include settlement and the development of mining and oil extraction industries and shipping across the Arctic Ocean. Limitations for human activity include damage to structures caused by freeze–thaw in the active layer and ground subsidence.

1.6 What are the methods used to manage glacial environments and how successful are these strategies?

Management of the impacts of glacial processes and landforms on human activity

Methods used to manage the impacts of glacial processes on human activity include prevention or control measures in the form of **soft** and/or **hard engineering** strategies. These include constructing strong, resistant buildings; constructing **avalanche barriers** on mountain slopes; and **planting trees** to break the flow of an avalanche. In response to the loss of life caused by the 1999 Galtür avalanche, a barrier wall was constructed. The choice of strategy is dependent on the nature of the human activity; the density of human settlement; the nature of the impact; the frequency of occurrence and degree of intensity of impact, loss of life and injury; and the damage caused to property and infrastructure.

Management of the impacts of human activities on glacial environments

Strategies used to manage the impacts of human activities on glacial environments include **prevention of access** to and **bans on use** of (aspects of) the area under impact. The occupation of land and character of land use in glacial environments can be controlled by means of **planning controls** and **zoning**, which affect access, location and design of buildings, and infrastructure. In Snowdonia National Park a variety of strategies, including repairing eroded upland footpaths, has been implemented to manage the impact of human activities in a formerly glacial environment.

Assessment of the success of strategies for managing either glacial processes/landforms or human activities

The assessment may include a financial cost–benefit analysis. This examines the extent to which strategy aims are achieved in terms of controlling or lessening undesired impact(s); the projected life of the management strategy or strategies; or the extent to which each involved agency or body feels that any implemented strategies have achieved objectives. Finally, having considered the interests of the different participants and the actual evidence 'on the ground', a summary evaluation of the success or lack of success of the strategy or strategies needs to be made.

Summary

After studying this topic, you should be able to:

- explain how a glacier operates as a system and how it responds to climate changes
- describe and explain the processes of glacial weathering and erosion and explain the development of named landforms associated with these processes
- describe and explain the processes of glacial transport and deposition and explain the development of named landforms associated with these processes

- describe and explain the processes and landscape changes associated with deglaciation
- understand that glacial environments are important as glacial processes and landforms impact on human activity and human activities impact on glacial environments
- know and understand a range of methods used to manage glacial environments and be able to assess the success of strategies used either to manage the impacts of glacial processes and landforms on human activities or the impacts of human activities on glacial environments

Theme 2(b) Coastal landforms and their management

1.1 What is a coastal system and what are the dynamics of coastal environments?

The coastal system

The coastal system is one of **inputs** and **outputs**. There are two systems:

- The **cliff system**, with *inputs* comprising the sub-aerial processes of **weathering** and the atmospheric process of wind erosion; a *throughput* comprising cliff **mass movement** of **falls**, **slips** and **slumps**, and an *output* of sediment at the base of the cliff, which is either deposited or transported by marine processes.
- The **beach system**, with an *input* of sediment from longshore drift, the cliff and offshore, a *throughput* of longshore drift and an *output* of longshore drift and destructive waves carrying sediment offshore.

Coastal sediment cells

Coastal **sediment cells** are areas of coast usually defined by headlands within which marine processes are largely confined, with limited transfer of sediment from one cell to another. They vary in size depending on the nature of the coast.

The state of dynamic equilibrium in the coastal system

The relationship between inputs and outputs is constantly changing — it is dynamic — and the system works towards an equilibrium position where inputs equal outputs. Erosion, transport and deposition occur, giving rise to the concept of **dynamic equilibrium**.

Wave types and characteristics and their variations over time and space

There are two extreme forms of wave: destructive and constructive waves. They have different characteristics and occur in different places determined by the local configuration of the coastline and/or prevailing wind conditions.

- **Constructive** waves are low, flat and gentle, with wavelengths up to 100 m and a low frequency of 6–8 waves per minute. They are characterised by a relatively more powerful swash, which carries sand and shingle up the beach, and a relatively weaker backwash. Constructive waves contribute to the formation of beach ridges and berms (Figure 5).

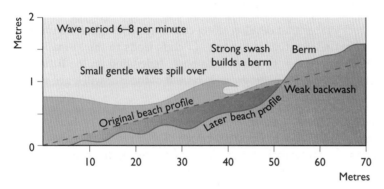

Figure 5 Constructive waves

- **Destructive** waves tend to occur during storms, are steep in form and break at a high frequency, at 13–15 waves per minute. They have a plunging motion that generates little swash and a relatively more powerful backwash; this transports sediment down the beach face, resulting in a net loss of material (Figure 6).

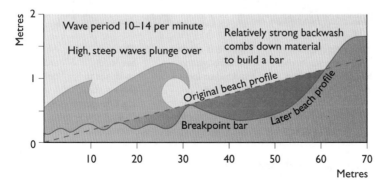

Figure 6 Destructive waves

19

Knowledge check 5

What does the term dynamic equilibrium mean as applied to a coastal environment?

Most beaches experience the alternating action of constructive waves in summer and destructive waves in winter, resulting in an annual cycle of beach growth and decay.

Wave refraction concentrates wave energy on headlands, which causes erosion, while dissipating energy in bays, which causes deposition.

1.2 What are the processes of coastal erosion and what are the resultant landforms?

Weathering and erosion in the coastal zone

Weathering in coastal environments includes physical disintegration caused by such processes as **freeze–thaw**, **salt crystallisation** and **wetting and drying**; chemical decomposition includes **solution** and **carbonation**. The variety of intertidal organic life encourages **biotic** weathering, with activity ranging from that of the roots of seaweed to acid secretions by limpets, barnacles and seagulls. Processes of erosion in coastal environments include **corrosion**, **hydraulic action**, **abrasion** and **attrition.**

Landforms of coastal erosion

Cliffs are rocky faces that develop along coastlines where marine undercutting has been ongoing. Undercutting at the base of cliffs leads to the formation of a **wave-cut notch**, then collapse by fall or slumping, depending on the **lithology** and **dip** of the rocks (see 1.4 below). Progressive retreat of the cliff will leave a gently sloping intertidal **wave-cut platform**.

Caves develop where lines of weakness such as joints or faults are exploited. **Arches**, e.g. Durdle Door in Dorset, form on headlands where caves erode back-to-back. **Stacks**, e.g. Old Harry, also in Dorset, result when the arch collapses. **Stumps**, e.g. Harry's Wife — located next to Old Harry — remain after the stack collapses (Figure 7).

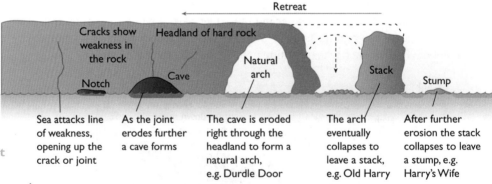

Figure 7 Cave–arch–stack–stump erosional sequence

Sea-level rises and erosion

Landforms created with rises of sea level are associated with either eustatic rise after a glacial retreat, as with the **Flandrian transgression**, or local areas

Eustatic adjustment
A worldwide change in sea level. Isostatic readjustment refers to a localised change in sea level due to changes such as the post-glacial uplift of land once the weight of the ice is removed.

of **subsidence**, as in the case of the south-east of England. Landform features include those associated with marine erosion and the submergence of the land by an encroaching sea, such as **rias** and **fjords**. For example, the **raised beaches** of Gower are wave-cut platforms eroded during a period with higher sea levels.

1.3 What are the processes of marine transport and deposition and what are the resultant landforms?

Transport and deposition in the coastal zone

Beach sediment is transported progressively along a beach by the process of **longshore drift**. This is caused by the oblique approach of waves and consequently swash, followed by the direct return to the sea of the backwash (Figure 8).

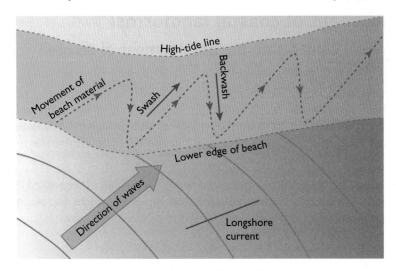

Figure 8 Longshore drift

The concentration of tidal flow produces tidal currents that are also important in transporting sediments.

Landforms of coastal deposition

Spits are banks of sand and shingle projecting from the shoreline into the sea. They need a supply of sediment from longshore drift to build and maintain them. Often the far end is **hooked**, being formed either by wave refraction or local wave approach from a different direction. **Double spits** occur where longshore drift extends one spit in the direction of the prevailing winds and another from the opposite direction, as at Poole in Dorset.

A **tombolo** is a spit joining an island to the mainland. For example, Chesil Beach links the Isle of Portland to the Dorset mainland.

Spits may grow across small bays, eventually closing them in with complete sand ridges called **barrier beaches**, e.g. Slapton Sands in Devon.

Examiner tip

Show the examiner your ability to make connections between your understanding of sea level rise and climate change (covered at AS, Unit G1). These connections are known as synoptic links and will be credited.

Bay-head beaches are areas of sand or shingle beach occupying part of a bay bounded by projecting headlands, e.g. Barafundle Bay in Pembrokeshire.

Offshore bars are ridges of sand and/or shingle developed offshore on a gently shelving coastline.

Cuspate forelands are triangular outgrowths of shingle ridges formed by longshore drift from opposite directions, e.g. Dungeness in Kent.

Sea-level rises and deposition

Sea-level rises will create **coastlines of submergence**. Eustatic rises in sea level are associated with glacial retreats and give rise to inundation of low-lying areas by the sea. **Estuaries** are created, which at low tide expose mudflats and deposited material.

1.4 What is the role of geology in the development of coastal landforms?

Lithological controls on the development of coastal landforms

Beach material is often made up of locally eroded rock whose character influences beach characteristics. Shingle, for example, can retain a higher slope angle than sand and allow greater infiltration. The headlands and bays on the Gower Peninsula are produced by alternate layers of resistant **carboniferous limestone** rocks and weaker **'Namurian' shales** (Figure 9).

Igneous rocks such as **granite** erode more slowly and produce steep-sided cliffs, e.g. Land's End. Cliffs composed of **unconsolidated sands** and **clays**, e.g. at Barton-on-Sea, Hampshire, suffer from the complex mass-movement processes of slides, slumps and flows.

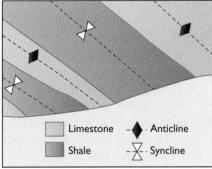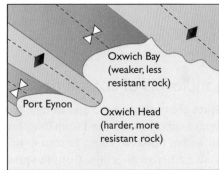

Figure 9 Headlands and bays (Gower Peninsula)

Structural controls on the development of landforms

Geological **structures** incorporating bedding planes, faults and cracks can add distinctive features to coastal cliff lines, such as the shape of caves and other local features like **blowholes** and **geos**.

The orientation of the coastline with the local geology is a very important factor (Figure 10). If the geological trend is **concordant**, i.e. parallel to the coast, then a **Dalmatian coastline** of coves and solid rock bars is created. Differential geology at right angles to the coast will result in a **discordant coastline** with bays and headlands, e.g. Isle of Purbeck, Dorset.

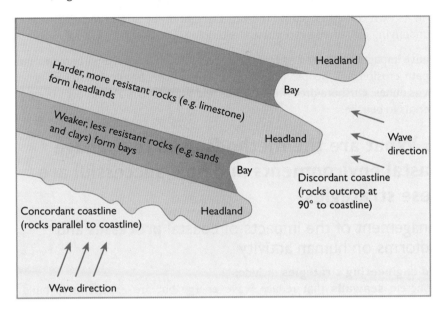

Figure 10 Concordant and discordant coastlines

1.5 Why do coastal environments need to be managed?

The impact of coastal processes and landforms on human activity

Coastal processes can impact on human activity in both a negative and positive way. Processes of weathering and mass movement along the coastline can endanger buildings, as with the collapse of the Holbeck Hall Hotel in North Yorkshire in 1993. Transport of material can silt up estuaries and make harbour entrances more difficult to navigate, e.g. Poole Harbour in Dorset, and the Newhaven ferry terminal on the River Ouse estuary in East Sussex. However, deposition also creates beaches that can be used for tourism, e.g. Studland Bay, Dorset.

Coastal landforms impact on human activity. Coastal lowlands are important for food production. Coastal ports play a leading role in world trade, and the coast has become a popular place for leisure, recreation and tourism.

The impact of human activities on coastal environments

Important large-scale industries that use a lot of space often have coastal locations, e.g. steelworks at Port Talbot, oil refineries at Milford Haven, gas terminals at Easington, Yorkshire, and defence installations at Plymouth.

Examiner tip

When discussing the impact of coastal processes and landforms on human activity always provide detail of specific coastal processes and landforms and give clear and developed examples of these impacts.

Seawalls, jetties and docks will be created to serve these installations and these will alter local tidal currents and influence local erosion and deposition patterns.

Tourism will have an impact on processes in cases where local authorities install seawalls, promenades and groynes to build sandy tourist beaches. In the Holbeck Hall Hotel case, for example, the building itself could have played an instrumental role in its own collapse because the extra weight of the structure could have made the underlying clay more susceptible to slumping.

Negative impacts of leisure and recreation activities on coastal environments include footpath erosion and people trampling on areas with fragile coastal ecosystems, such as dunes. Offshore dredging of sands and gravels may also affect the supply of materials to beaches.

1.6 What are the methods used to manage coastal environments and how successful are these strategies?

Management of the impacts of coastal processes and landforms on human activity

Hard engineering strategies include:
- Concrete **seawalls** that reduce wave energy but are expensive to build and maintain.
- **Rock armour (rip-rap)**, consisting of large boulders of hard rock placed in front of seawalls and sometimes used as groynes. These are both expensive and ugly.
- **Groynes**, which are walls built at right-angles to the coast. Traditionally they have been made from wooden railway sleepers, but they can be made of rock armour. They prevent beach migration.
- **Gabions**, which are blocks made by wire-netting together medium-sized pieces of hard rock. These are expensive and can be ugly.
- **Revetments**, which are slatted and angled low wooden walls built parallel to the beach. They act to absorb wave energy and protect soft cliffs. They are ugly and liable to rapid damage.

Cliff regrading does limit slumping, particularly when combined with cliff **drainage schemes**, but this does not stop the loss of beach material.

Soft engineering strategies include:
- **Beach nourishment**, which is the artificial input of beach material to compensate for natural losses. Miami Beach in Florida is managed in this way, but such a strategy is extremely expensive.

The Holderness coast, in the East Riding of Yorkshire, is in active retreat and coastal management strategies are critical. The village of Mappleton is threatened by coastal erosion and its access road is within 50 m of the cliff edge. In 1991, **hard engineering** was undertaken to slow down the rate of erosion. A **groyne** made from granite boulders was built out to sea, behind which sand was trapped to form a beach. Another groyne and a **revetment** were constructed south of the first groyne and at right-angles to it to absorb the destructive energy of waves and restrict the

Examiner tip
Ensure that you understand the various hard and soft engineering strategies in terms of controlling or reducing undesired impacts in coastal locations.

Cliff regrading
Restructuring a cliff face to remove the steep gradient.

removal of beach material by longshore drift. There has also been an attempt to stabilise the boulder-clay cliffs by **regrading** the cliff face so as to reduce its angle and thereby reduce mass movement (Figure 11).

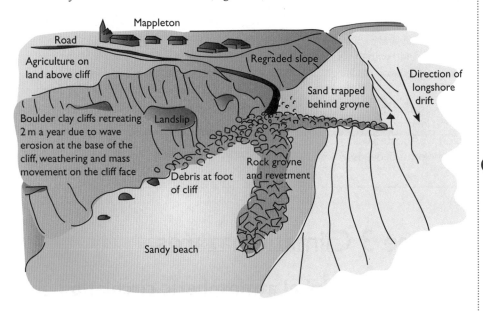

Figure 11 Coastal management strategies at Mappleton, Holderness

Knowledge check 6

Explain the difference between hard and soft engineering approaches to coastal management.

Management of the impacts of human activities on coastal environments

Lulworth Cove in Dorset receives around 750,000 visitors every year. However, there are many problems associated with tourism, including large car parks needed for vehicles, unsuitable or unsightly tourist shops, footpath erosion, noise and air pollution, rubbish and sewage, and erosion of the Geological SSSI associated with fossil-hunting and field trips.

Management of these human activities includes footpath management; information for tourists, such as display boards, leaflets, guided walks and talks and educational displays; the screening-off of 'eyesore' sites; and the setting-up of a coastal volunteer system.

Assessment of the success of strategies for managing either coastal processes/landforms or human activities

The management strategies put in place at Mappleton have reduced coastal erosion but not solved the problem. Local people are demanding more protection but the government and local authority are reluctant to undertake more work as they believe the costs of further coastal management outweigh the benefits. Additional work would only slow down, not stop, further coastal retreat. Also, trapping more sediment at Mappleton would create more erosion further south, because less sediment would be transported by longshore drift to build up beaches and protect the coast.

Summary

After studying this topic, you should be able to:

- explain how the coastal system operates and understand the dynamics of coastal environments

- describe and explain the processes of coastal weathering and erosion and identify and explain the development of landforms associated with these processes, including landforms of erosion resulting from sea level rise

- describe and explain the processes of marine transport and deposition and identify and explain the development of landforms associated with these processes, including landforms of deposition resulting from sea level rise

- describe and explain the effect of lithology and structure on the development of coastal landforms

- understand that coastal environments need to be managed as coastal processes and landforms impact on human activities and human activities impact on coastal environments

- know and understand a range of methods used to manage coastal environments and be able to assess the success of strategies used either to manage the impacts of coastal processes and landforms on human activities or the impacts of human activities on coastal environments

Theme 3 Climatic hazards

1.1 How does global atmospheric circulation give rise to global climatic zones?

Atmospheric movement

Solar energy (**insolation**) powers the atmospheric system and the energy circulations within it. The amount of solar energy (**heat budget**) received varies with latitude. The tropics have an **energy surplus** as they gain more from insolation than is lost by **radiation**. The higher temperate and polar latitudes have an **energy deficiency**, losing more by radiation than is gained by insolation. This imbalance in energy distribution sets up a transfer of heat energy from the tropics to higher latitudes.

The tricellular model: the Hadley, Ferrel and polar cells

This global transfer of energy is the basis of global atmospheric circulations, which give rise to the low- and high-pressure belts and the planetary wind systems associated with the Earth's three major convection cells: the **Hadley**, **Ferrel** and **Polar cells**. These make up the **tricellular** model that controls atmospheric movements and the redistribution of heat energy (Figure 12).

Knowledge check 7

Identify the main global locations of the low-pressure and high-pressure areas associated with the Hadley cell.

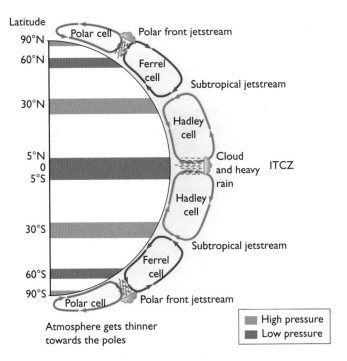

Figure 12 Convection cells and pressure belts

The patterns of winds and the world's pressure belts

This is illustrated by Figure 13.

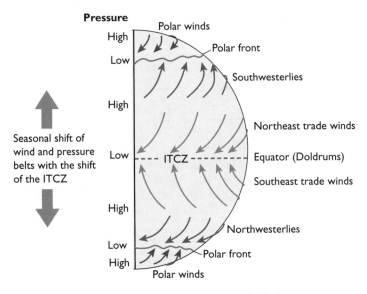

Figure 13 Pressure belts and associated wind systems

1.2 Why do seasonal and periodic variations of climate occur?

Seasonal variations

The reasons for seasonal variations in climate are:

- The seasonal movement of the **Inter-Tropical Convergence Zone (ITCZ)**, and pressure and wind belts associated with the movement of the sun's overhead position during the year.
- The effects of the warm and cool ocean currents.
- Temperature differences between continental landmasses and neighbouring ocean waters.

In the tropical region seasonal changes are far more marked in **savanna** and **monsoon** climates. In the temperate region seasonal changes are more pronounced for the **continental interior** and **east-coast margin**.

The following is an explanation of seasonal variations in **monsoon** climates.

Monsoon climatic type

Such climates occur mainly on the eastern side of continental landmasses in the tropics, extending across approximately 5–20 degrees of latitude.

The type is marked by a *distinct hot wet and a cooler dry season*, determined by the annual movement of the ITCZ between the tropics and associated movement of pressure belts and seasonal reversal of winds consequent on this. The monsoon climate regime is most clearly seen in the Indian subcontinent, but exists in other regions north and south of the Equator on the eastern edge of continents, e.g. east Africa.

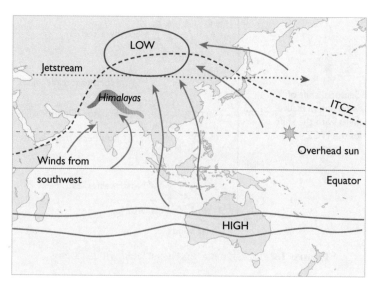

Figure 14 The wet monsoon season (June to October)

Examiner tip

Make sure that you understand the seasonal variations in at least one temperate or tropical climatic type.

The *wet monsoon season* (Figure 14) occurs with the movement of the ITCZ into the region. This brings an area of low pressure and draws in hot, moist winds from the ocean. Rainfall is increased by **orographic uplift**, where these moist winds are drawn over uplands, e.g. the Western Ghats in India. Temperatures are high, averaging 30°C. Humidity is also very high, with average rainfall around 2000 mm, decreasing with distance inland. Cyclones and hurricanes are frequent towards the end of the rainy season.

The *cooler dry season* (Figure 15) coincides with the extension of continental high pressure as the ITCZ moves back towards the Equator and across into the tropics beyond. With high pressure dominating, there is air subsidence and the outblowing winds are dry. Temperatures remain relatively high, at 25°C in lowland areas, and evaporation rates are also high. The weather is much more severe in mountain areas.

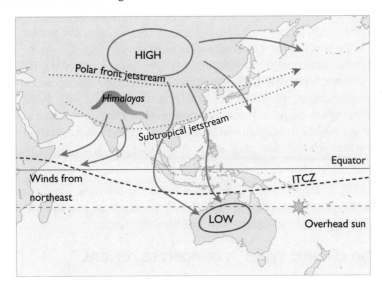

Figure 15 The cooler dry season (November to May)

Periodic changes in climate

Periodic variations in climate occur over both the long term and short term. **Glacials** and **interglacials** are examples of long-term changes. **El Niño/La Niña** cycles are examples of short-term changes.

1.3 What are the world's major climates?

For the examination, you need to have a broad knowledge and understanding of the world's major climates, including the distribution of the main climate types for the tropical and temperate latitudinal belts. However, detailed reference needs to be made to only ONE climatic type chosen from either a **tropical** or **temperate** region.

The main climatic types in tropical regions

The main influences on the climates of **tropical** regions are:

- The overhead or near-overhead position of the sun giving high insolation throughout the year.

Examiner tip

Show the examiner that you understand that climate change occurs on different time scales, both short-term and long-term. These temporal patterns of climate change were covered at AS in Unit G1.

- The position and seasonal movement of the ITCZ together with the tropical pressure belts' wind systems.
- The path of the upper jetstreams affecting the path of low-pressure systems.
- Differential heating of landmasses and oceans in the tropics affecting air-pressure patterns and seasonal wind directions.
- The effects of offshore cold currents on western land margins and warm currents on eastern margins.
- The position of mountain ranges and their effects on incoming moist winds off the ocean.

Savanna climatic type

5–20 degrees latitude either side of the Equatorial belt.

High temperatures of 35–25°C prevail all year because insolation is high. This climate type is distinguished by having a *hot, wet* season and a *cooler, dry* season. Humidity is highest in the wet season, but evaporation rates remain high during the cooler dry season. Rainfall occurrence is associated with the movement of the ITCZ towards the tropic in association with the apparent movement of the sun's position overhead. During the hot season, as this occurs, low pressure prevails with moist inblowing winds and rising air currents leading to **convection** rainfall. Rainfall amounts are most reliable towards the Equatorial latitudes, where they average 800 mm a year, but become less reliable towards the hot desert margins, where they average 300–400 mm annually.

The cooler, dry season in the savanna belt occurs at the time when high pressure and dry, outblowing winds prevail; this is when the overhead sun and the ITCZ move away to extend beyond the Equator towards the other tropic.

The main climatic types in temperate regions

The main influences on climate in **temperate** regions are:
- Mid-latitude position.
- The influence of the mid-latitude low-pressure belt and the atmospheric conditions along the polar front as well as the influence of the upper jetstream, except for continental areas in winter.
- The seasonal shift of the pressure and wind belts.
- The position and interaction at the margins of the different air masses affecting areas in temperate latitudes.
- Differential heating of the continental interior and ocean margins.
- The effects of ocean currents and the air above them.
- The location of upland ranges in relation to prevailing winds.

Maritime west margin European climatic type

Mid latitude, 35 to 55 degrees.

This climatic type is characterised by relatively mild temperatures (average seasonal range 5–20°C), along with high humidity and precipitation (averaging 600 mm) throughout the year. However, precipitation totals are significantly higher over upland areas in the face of prevailing moist westerly winds coming off the ocean,

Knowledge check 8

What characteristic is the key feature of savanna climates?

Knowledge check 9

What is convection rainfall? Name two other types of rainfall.

e.g. in the Cambrian Mountains of Wales. Conversely, precipitation totals are low in rain-shadow areas, e.g. lowland East Anglia.

The temperatures and precipitation figures are mainly influenced by the mid-latitude position, low-pressure belt and the mild westerly prevailing winds. The latter are warmed by warm currents, e.g. the Gulf Stream, on the west margin of landmasses.

The weather is strongly influenced by the position of the polar front, the associated jetstream, and the passage of westerly-moving depressions along the front, with intervening spells of anticyclonic conditions. These are linked to the position and extent of the main air masses influencing the continental west margins in mid latitudes: the **Polar Continental**, **Polar Maritime**, **Arctic Maritime**, **Tropical Maritime** and **Tropical Continental** Air Masses. The interaction between these air masses, together with the associated upper jetstream and Rossby waves, influence the occurrence and development of depressions along the polar front.

Persistence of one of the continental air masses across these western margins can bring long spells of dry summer weather, but in winter 'anticyclonic gloom' conditions may occur. In contrast, the passage over the area of a deep, fast-moving depression can bring storm conditions with gale-force winds and heavy rainfall.

Examiner tip

Learn climatic data for your chosen climate type to exemplify its characteristics of rainfall amount and distribution and variations in temperature at a diurnal and seasonal level. Use these figures in your answers to make them more rigorous and detailed.

1.4 What are the causes of low pressure and high pressure hazards?

The role of jetstreams and Rossby waves in controlling the formation of weather systems

Jetstreams and **Rossby waves** control the formation of weather systems. Between the different atmospheric cells, at a height of about 5 miles, within the **tropopause**, are the jetstreams (Figure 12): the **polar jetstream** (40–60°N+S) and the **subtropical jetstream** (25–30°N+S). These jetstreams move air at high speed (up to 130 mph) horizontally around the Earth and give rise to Rossby waves. The number of waves varies throughout the year but in summer there are usually between 4 and 6 waves. In winter, there are 3 waves.

In the mid latitudes of the northern hemisphere, Rossby waves are connected with the formation of **depressions** and **anticyclones**. As air travels west to east into a trough, it slows down and piles up and causes **convergence**. Convergence in the upper air causes a downflow to the ground, creating high-pressure systems at ground level. As air leaves the trough it speeds up and diverges ahead of the next trough. **Divergence** in the upper air causes low-pressure systems at ground level. At times the waves are few and shallow, giving a **high zonal index** and a succession of low-pressure systems. At other times the flow becomes more pronounced, giving a **low zonal index** and causing the formation of blocking, high-pressure systems.

Low-pressure system formation and associated hazards of storms, tropical cyclones and tornadoes

In the tropics, hazards associated with low-pressure systems are **tropical storms** and **cyclones** with torrential rain and high winds. These hazard conditions, usually

Climatic hazard An 'extreme climatic/weather event causing harm and damage to people, property, infrastructure and land uses'. It includes not only the direct impacts of the climate/weather event itself but also the other (secondary) hazards 'triggered' by that event — e.g. landslides triggered by torrential rain.

created towards the end of the hot season (August–November in the northern hemisphere), are generated in exceptionally deep, fast-moving depressions over oceans off the east margins of continents in the tropics and subtropics. They trigger the secondary hazards of **flooding**, **storm surges** and **sea incursions**, **landslides**, **mudflows** and **windborne debris**.

In the temperate region, hazards associated with low-pressure systems include **severe storms**, **heavy rainfall** or **snowfall** and **gale-force winds**. These conditions are generated in exceptionally deep and fast-moving depressions, which are most likely to occur in autumn and spring along the polar front. They trigger the secondary hazards of flooding, sea incursions (especially where the deep depression coincides with a time of very high tides), landslides and windborne debris.

Tornadoes are small cells of very low atmospheric pressure formed where warm, damp air meets cool air from continental interiors. Tornadoes are associated with high wind speeds and cause structural damage to buildings along their narrow storm path.

High-pressure systems and associated hazards of drought in tropical climates, or drought, frost and fog in temperate climates

Examiner tip
You should show knowledge and understanding of the climatic causes of, and the weather associated with, low- and high-pressure systems.

In the **tropical** climates, the hazards associated with high-pressure systems are low rainfall, high evaporation rates and **drought**. These trigger the secondary hazards of **falling water tables**, **loss of vegetation**, **wildfires**, **soil erosion** and associated **desertification**. These hazards are associated with anticyclonic conditions, which are due to the continued persistence of the subtropical high pressure over continental areas. This limits the ITCZ zone to lower latitudes (nearer the Equator) than is normal for the time of the year. **Global warming** is a further contributory factor, exacerbated by people's misuse of their environment.

Examiner tip
You should show knowledge and understanding of the human circumstances that constitute the hazard: the path/track/spatial extent of the weather event; the density/distribution of the people; the types of human activity in the areas affected; and the preparedness of the authorities/people to cope.

In **temperate** climates, the hazards associated with high-pressure systems include **drought** in summer and **frost** and **fog** in winter. They may trigger secondary hazards in summer: **falling water tables** and **loss of vegetation**; and in winter: **temperature inversion** with air pollution intensifying the fog conditions. These conditions are associated with a persistent stationary anticyclone, which in summer is usually associated with the extension into higher latitudes of the subtropical high pressure. In winter the conditions are usually associated with the extension of the continental high pressure towards the coastal margin of the landmasses.

1.5 What are the inter-relationships between human activity and climate?

The short-term and long-term effects of low-pressure climatic hazards on human activity

Low-pressure climatic hazards have both short-term and long-term effects on human activity. You should study these with reference to at least one specific low-pressure event in EITHER a tropical OR temperate climate (for an example, see Table 1).

Table 1 The impacts of Cyclone Nargis (May 2008) on Burma (Myanmar)

Economic impacts	Social impacts	Environmental impacts
• Rice crops destroyed • Damage estimated at over US$10 billion • Fisheries damaged	• 130,000 deaths • 75% of health centres damaged • 800,000 homes destroyed • 260,000 moved into refugee camps • Food shortages • Contaminated drinking water resulted in fever and diarrhoea	The environment and natural ressources underpin the livelihoods of people in Myanmar: • Damage to 35,000 ha of natural and planted mangroves and other trees • Pollution of surface and groundwater • Salination and erosion of agricultural lands — 63% of paddy fields flooded by storm surge • Sedimentation of rivers

Examiner tip

When assessing the effects of low-pressure climatic hazards, a decision as to whether the short-term impacts are greater than the long-term impacts or whether the economic, social or environmental effects are the most important, can form part of your assessment.

The short-term and long-term effects of high-pressure climatic hazards on human activity

The hazards associated with high pressure have both short-term and long-term effects. You should study these with reference to at least one specific high-pressure event in EITHER a tropical OR a temperate climate. For example, you could study the effects of drought in Australia (for further information visit: **www.bom.gov.au/climate/drought/livedrought.shtml**).

Examiner tip

Try to provide examples of the effects of hazards associated with recent climatic events to make your answers as topical as possible.

The impacts of human activity on climate in both the short and long term

Review your understanding of how human activity affects climate change, which was covered at AS in Unit G1.

1.6 What strategies are used to reduce the impact of climatic hazards?

Strategies to reduce the impact of hazards associated with low-pressure and high-pressure systems include **monitoring**, **prediction** and **warning** of future hazards, **immediate response** to lessen the impact once the hazard has occurred, and **long-term planning**.

Strategies to reduce the impact of low-pressure climatic hazards

In the case of Hurricane Katrina, which struck the Gulf coast of the USA in 2005, hurricane warnings were given and emergency services were in place but the strategies were not successful due to the failure of the levees to protect the city of New Orleans and due to the slow response of the US Federal Emergency Management Agency (FEMA). LEDCs are usually less prepared and often depend on aid from MEDCs. Other strategies used in poorer countries include building hurricane shelters, strengthening and raising embankments, planting mangrove trees to absorb storm surges and educating people about the risk.

Strategies to reduce the impact of high-pressure climatic hazards

In the case of drought in southeast Australia, sustainable solutions include water saving, water re-use and water treatment.

Strategies to reduce the impact of human activity on climate

Review your understanding of strategies to reduce the impact of human activity on climate, which was covered at AS in Unit G1.

Summary

After studying this topic, you should be able to:

- explain how global circulation gives rise to global climatic zones
- describe how and explain why seasonal and periodic variations of climate occur
- have a broad knowledge and understanding of the main climatic types in tropical and temperate regions, but have detailed knowledge and understanding of one climatic type chosen from either a tropical or a temperate region
- explain the causes of the hazards associated with low-pressure and high-pressure systems

- know and understand the short-term and long-term effects of both low-pressure and high-pressure climatic hazards on human activity and the impacts of human activity on climate in the short and long term
- know and understand strategies to reduce the impact of hazards associated with low-pressure and high-pressure systems as well as strategies to reduce the impact of human activity on climate
- assess the success of strategies for reducing the effects of either low-pressure or high-pressure climatic hazards

Theme 4 Development

1.1 What is development and what is the development gap?

Changing definitions of development

'Development' is very difficult to define, partly because the definition is dynamic. It is self-evident that different countries throughout the world are at different stages of development, but different groups of people mean different things by 'development'. A working definition is *'an increase in standards of living and quality of life for an increasing proportion of the population'*.

Earlier views on 'development' emphasised economic expansion and increased output, but the current view is much broader, involving social and cultural advancement as well as technological change and economic growth — and more recently, sustainable development.

Conventional development divides

Until the late 1980s countries were classified as belonging to either the 'First', 'Second' or 'Third' Worlds. The **First World** referred to countries that developed on the basis

Sustainable development
Development that meets the needs of the present, without compromising the ability of future generations to meet their own needs.

WJEC A2 Geography

of capitalism, whereas the countries of the **Second World** developed on the basis of a command economy. The idea behind these terms was that each 'world' represented a path to development and **Third World** nations could choose between the first way (capitalism), a second way (communism), or invent a third way.

'**Developed**' is a term usually applied to a country or region that has a high standard of living and an advanced economy based on the effective utilisation of resources. Countries like this are often referred to as 'More Economically Developed Countries' (**MEDCs**), a term which acknowledges the fact that those countries that are not 'developed' in the sense of this definition may be developed in other non-economic ways, such as with regard to cultural, religious or social conditions. '**Developing**' is a term usually applied to a relatively poor country or region that has a low standard of living but which is beginning to achieve some economic and social development. In contrast to MEDCs, countries with low levels of economic development are referred to as 'Less Economically Developed Countries' (**LEDCs**).

The **Brandt Report**, published in 1980, highlighted the growing gap in social and economic development between the 'developed' countries of the world, **The North**, and the 'less-developed' countries, **The South**. However, the North/South classification is simplistic and distorts a geographer's view of 'north' and 'south'.

The development gap

It is clear that there are groups of countries that share common characteristics. Some of these groupings are very polarised, leading to ideas of a **development gap**.

The development continuum

While it is true that massive contrasts occur, there are always countries that score at intermediate levels, leading some to conclude that there is a development continuum rather than a gap. The idea of a continuum tends to obscure the extent of extremes and many prefer to continue to use the term 'gap'.

1.2 How can development be measured and how useful are these measures?

Simple and composite indicators used to measure development

Many indicators may be used to quantify the level of 'development' of a country or region.

- **Simple** indicators measure only one aspect of development and include **GDP**, **GNP** (now GNI), **adult literacy**, **life expectancy**, **daily calorie supply**, **infant mortality**, **cars per 1000 people**, **percentage of employment in agriculture, manufacturing and services** and even the **Big Mac index**.
- **Composite** indicators are more comprehensive as they measure more than one aspect of development. For example, the **Human Development Index (HDI)** gives a country a score from 0 to 1 using three variables based on adjusted income, education and life expectancy. On this basis Norway is the most developed nation, with an HDI of 0.938, and Zimbabwe is the least developed with an HDI of 0.140 (2010 statistics).

Examiner tip
You need to appreciate that Brandt's 'North' and 'South' do not correspond with the northern and southern hemispheres. Many countries in Brandt's 'South' are located in the northern hemisphere.

Knowledge check 10
Rank groups of countries from the most developed (MEDCs) to the least developed (LDCs).

Development continuum The variation in the economic development of countries.

GNI per capita in US dollars The sum of all goods and services produced in a country plus taxes and income from abroad, divided by the population. GNI was previously known as GNP.

Knowledge check 11
Why might a country with a particular HDI rank have a different rank on the GDP or per capita GNP scale?

Qualitative indicators

By **qualitative indicators** of development we mean aspects of development that may not be easily quantifiable. Qualitative indicators have been developed due to the recent emphasis on measuring development in terms of issues, such as **freedom**, **security** and **sustainability**, rather than by **statistics**. Qualitative indicators may be more problematic but reflect more accurately the ways in which development is now viewed.

The use of both **quantitative** and **qualitative** indicators is necessary in assessing the level of development of a given country and together they can yield important and unexpected insights into development that neither quantitative nor qualitative indicators could generate on their own.

Limitations of indicators

A problem with many indicators is that they can be misleading because most are averages and therefore do not show how far the benefits of development are shared within a population. Another problem with most measures of development is that they do not show the harmful side-effects that can occur. For instance, a rise in car ownership indicates a general rise in living standards, but some of the population will suffer from increased noise and air pollution and may take the view that their living standards have fallen. As with all statistics, development indicators are sometimes incomplete or inaccurate. Also, for some countries data are not reliable.

1.3 What factors have led to contemporary differences in development?

Physical, economic, social, political and cultural factors affecting the rate and nature of development

Frank's dependency theory suggests that developed countries control and exploit less-developed countries. This produces a relationship of dominance and dependency, possibly leading to poverty and underdevelopment in LEDCs. **Rostow** argues that all countries have the potential to break the cycle of poverty and develop through 5 stages of economic growth.

Within global regions, the level of development is often similar. Most European countries are well developed; most African countries are at low development levels and many parts of Asia are developing rapidly. But within regions of the world there are also variations dependent on resource endowment, government policy, and a wide range of other factors.

Malaysia has developed as a successful, second-generation, **newly industrialised country (NIC)** due to tax revenues from the **physical** resource of oil in the South China Sea, which provided initial funds for the country's development.

Political factors also contributed, including careful state planning with the government, led by Dr Mahatir Mohamad, managing the economy and controlling **foreign direct investment (FDI)**.

Knowledge check 12

What is foreign direct investment (FDI)?

Economic factors were also significant, including not just the taxes from oil, but also low labour costs (11% of labour costs in the USA) and laws that favoured investors. Other contributory economic factors included the growth of export-led industries in economic priority zones (EPZs), free investment zones (FIZs), and free trade zones (FTZs).

Social factors, such as state investment in higher education, and **cultural** factors, including the 'Look East' policy and 2020 Vision (Malaysia hopes to achieve developed-world status by 2020) also help to explain the country's rapid rate of development.

The globalisation of economic activity and the rise of NICs/ RICs and oil-rich countries

The opportunity to develop and the rate at which development has been taking place have been much influenced by the globalisation of the world economy. The biggest impact of this has been the **outsourcing** of manufacturing and services from developed countries to other parts of the world. This has, in turn, encouraged home-grown manufacturing in areas surrounding the focus of the outsourced activity. Tertiary activity has also moved out. Much of this has been low-level call-centre work, but higher-end activities, such as software design, have also moved.

Globalisation The generic term for the process of integration in the realms of trade, economic relations and finance.

Greater economic integration, such as that between Mexico and the USA, has stimulated development. Huge reserves of money generated by newly industrialised economies have made capital available to stimulate the establishment of new economic activities. The increased scale of the world economy has stimulated the extraction of raw materials and utilisation of energy sources, increasing their prices and injecting income into economies that had previously shown little sign of growth. However, it would be wrong to class the entire developing world as exactly the same. Differences exist within the countries and between the countries.

Two significant groups of countries are the **oil-rich nations** and the **newly industrialised countries (NICs)**.

Knowledge check 13

What does the term OPEC stand for? Name two members of OPEC.

The 1970s saw huge increases in the price of oil as Arab countries withheld supply. This meant that any developing country with oil suddenly had a source of great wealth. Those without it faced a major disadvantage as they too had to pay the inflated price for fuel. The consequence for countries with oil was that they could now invest money in trying to alleviate poverty and promote education and health. Many also invested in industries that refined oil, adding great value to their export. Additional profits were invested in overseas ventures or loaned to other countries. As a consequence many of these countries, such as Saudi Arabia, amassed great wealth.

Newly industrialised countries are those that have recently seen substantial growth in their manufacturing output and, as a consequence, growth in their exports. They include South Korea, Singapore and Taiwan, as well as the territory of Hong Kong. Many have followed a similar system. Firstly, the country invests in industries that can produce goods that the country would normally import and supports these new industries by putting extra taxes on imported goods to make them uncompetitive. Then, when these industries are established, the countries look to replicate many of the products in the world export market. They concentrate on high-technology industries, first copying existing products and then improving them. The economy of such a country typically grows by 6–8% a year.

South Korea, for example, took advantage of its links with the USA. The USA, Japan and Europe provided the country with significant aid payments that it invested in iron, steel, shipbuilding, textiles and chemicals so it no longer had to import these. Once these were established South Korea invested in export-led industries. Malaysia, India and China are examples of **recently industrialised countries (RICs)**.

1.4 How and why are development patterns changing?

Contemporary global development divisions

The World Bank classifies economies by their gross national income (GNI) per capita. There are 209 economies, classed as **low-income**, **lower-middle income**, **upper-middle income** or **high-income**. The classification is updated every year to keep pace with change. Of the 35 nations classified as low-income countries, 26 are located in sub-Saharan Africa.

Development issues in a changing world to include sustainable development and the status of women

The concept of **sustainable development** dates from the first Global Environmental Summit (1972), held in Stockholm. The concept was developed in the Bruntland Report (1987), where it was defined as *'development which meets the needs of the present without compromising the ability of future generations to meet their own needs'*. Although aspects of sustainable development can be measured, it is difficult to quantify sustainable development comprehensively, and also we do not know what the specific needs of future generations will be. However, for a country to be considered 'developed', it should be concerned about its environment and long-term sustainability. In this respect the USA cannot be considered developed in sustainable terms as it is responsible for the emission of 25% of the world's carbon.

Gender inequality is often a barrier to development. If only half the population can gain from development, any benefit is diluted across the whole of society, making the overall impact less. When Bill Gates was asked whether it was realistic for Saudi Arabia to be in the top 10 countries in the world in technology by 2010 he replied that it would not get close, as the country is not fully utilising half the talent.

Economic change resulting in differentiation between different groups of countries

Economic change has affected world patterns of development, necessitating significant alterations to the 'three-worlds' model and Brandt's 'North/South divide'. Some countries have declined or stagnated — such as Tajikistan in central Asia, since the breakup of the former Soviet Union, of which it was a member. Other countries have developed beyond all expectations, such as Malaysia. As a result the **Third World** or **South** has become substantially differentiated.

Examiner tip

Learn up-to-date GNI statistics data for a country in each of the four World Bank divisions to illustrate contemporary development divisions.

Examiner tip

When referring to Africa you need to be aware that it is a continent made up of 53 countries including the island states, with a diverse range of physical, economic and political characteristics.

1.5 What hinders the closing of the development gap?

The burden of Third World debt

The greatest obstacle to development for an individual country is **indebtedness**. Countries that were at a low level of development in the past were loaned money through the World Bank and International Monetary Fund. Any money that the country generated had to be spent on paying interest on the loan before repaying the debt, and reinvestment in the economy was impossible. Such countries became caught in a poverty trap. They became what were defined as **'Heavily Indebted Poor Countries' (HIPCs)**. Special arrangements to relieve this debt have been developed by richer nations, such as the Multilateral Debt Relief Initiative (MDRI), but many believe this is still not enough to allow real development to take place.

> **Knowledge check 14**
>
> Can you name a country or countries that have qualified for HIPC Initiative Assistance?

Trade blocs

Countries have often come together to create trading blocs (e.g. the **European Union**), which greatly benefits each member of the bloc. However, countries outside the bloc may face quotas or tariffs that make it almost impossible to sell the commodities they have to offer. Within blocs, regulations often make it possible for producers, particularly of food crops, to generate huge surpluses. The surpluses are then sold at below cost price on world markets (dumping). Countries outside the blocs find that the commodities being 'dumped' are the same ones they themselves have to sell, but the dumped commodities are sold at below the price that could give the outsider countries any profit. This means the latter countries' only means to economic development is undermined.

> **Knowledge check 15**
>
> Can you name two trade blocs, other than the EU?

Social constraints and cultural barriers

Gender inequality is a barrier to development. However, the **caste system** in India can also be viewed as a significant additional, cultural constraint. Hinduism is deeply rooted in India's culture, particularly through the caste system, which discriminates against the lowest caste, the **Untouchables** or *Dalits*. The Indian government has tried to reduce discrimination by ensuring that a percentage of public-sector jobs are reserved for certain subcastes of Dalits. Research in coastal villages in the Indian state of Andhra Pradesh showed that caste was a key factor influencing access to public services and facilities.

1.6 What types of strategies exist for reducing the development gap and how effective are these strategies?

Different types of aid: bilateral, multilateral and emergency aid

One way to try to overcome a low level of development is by accepting aid. However, not all aid is given with the intention of stimulating development. **Emergency aid** is directed towards providing relief after a disaster. One country may help another directly (**bilateral aid**) — often when a former colonial power assists a former colony after independence. More widely based assistance is usually termed **multilateral aid**. The aid may be monetary, but can be in the form of equipment, education programmes and/or expertise. If such aid is specifically directed at promoting development it is often termed **structural aid**.

Many criticisms have been made of aid. Schumacher (1974) said aid was a process by which *'you collect money from poor people in rich countries and give it to rich people in poor countries'*. To be successful, aid must be effective, transparent and sustainable. Useful examples include education (particularly of girls), safe-water and sanitation projects, the provision of small loans (of less than US$200) to poor people, the eradication of diseases, and research to develop new crop varieties resistant to viruses and drought.

Free trade and fair trade

A point that has been noted is the way that economic blocs can hinder trade that could generate income and development. Removing barriers and promoting trade can allow development to take place. The **World Trade Organization (WTO)** is committed to increasing **free trade**. However, the governments of countries with well developed economies are often reluctant to give up their position of advantage, and removing barriers is not as simple as it sounds. There are campaigns to promote **fair trade**, whereby producers are paid a reasonable price even when it is possible to drive the price down. Such trade encourages safe and non-exploitative working conditions, and avoids the use of child labour.

Foreign direct investment and the role of transnational corporations

Development has occurred extensively when manufacturing industry is introduced or opportunities to exploit valuable resources are taken. Governments of countries lacking development often work hard to encourage overseas-based companies to set up assembly or manufacturing plants or to mine resources. This may be arranged between governments but has most impact when a transnational corporation (TNC) is attracted and establishes a large plant or mine. Other companies that provide raw materials, services, transport and communications are also attracted and this often promotes rapid development.

The impact of TNC mining activity in Botswana has been very positive, with the country's per capita GNP increasing from US$800 in 1975 to US$16,500 in 2007. This has mostly been as a result of the mining of diamonds.

However, there can be a serious downside if a TNC moves on before a self-perpetuating multiplier has been established. Very often pollution occurs, and traditional attitudes and values come under great strain.

Debt rescheduling, debt abolition and debt-for-conservation swaps

As debt has been such a massive handicap to development, ways of eliminating debt have become important. **Rescheduling** can make repayment easier, but there is a

Examiner tip
Avoid confusing free trade with fair trade. Free trade exists when trade between countries is not restricted in any way by tariffs, quotas or other barriers. Fair trade is a social movement and market-based approach that aims to improve trading conditions for producers in developing countries.

Knowledge check 16
Which world region has benefited the least from the growth in free trade over the last two decades?

Transnational (TNC) or multinational company (MNC)
A company that operates in more than one country.

strong lobby for **debt abolition**. The HIPC Initiative currently identifies 41 countries, most of them in sub-Saharan Africa, as potentially eligible to receive debt relief. One solution that also sets out to achieve a further objective is a **debt-for-conservation** arrangement. Here, a debt, or a portion of it, is written off and the money that would have been paid in interest on the debt is used to pay for conservation measures. In Peru, ten of the most biologically diverse yet critically endangered rainforest areas escaped deforestation thanks to a landmark debt-for-conservation swap that the US Nature Conservancy helped facilitate in June 2002.

The UN has set targets for development through the eight 2015 **Millennium Development Goals**. The setting of these goals has had an excellent result in terms of highlighting issues, but success in achieving them has so far been limited.

One example of a narrowing of the development gap is that of Vietnam. Since the 1980s, FDI, improvements in trade — with membership of the Association of Southeast Asian Nations (ASEAN) bloc in 1995 and the WTO in 2006 — together with aid (e.g. £50 million a year from the UK Department for International Development) have operated to improve human development indicators and economic growth rates. Poverty fell from 58% in 1992 to 12% in 2010, life expectancy has improved, adult literacy is rising and primary-school enrolment rates are 97.5%. In the last 20 years, Vietnam reached an average economic growth rate of 7.5%. However, the country still faces challenges such as inequality, corruption, bureaucracy and environmental deterioration.

Knowledge check 17

What progress has been made in sub-Saharan African countries towards the Millennium Development Goals (MDGs)?

After studying this topic, you should be able to:

- define the terms 'development', 'development gap' and 'development continuum' and understand how the definition of development has changed
- know and understand a range of simple, composite and qualitative indicators that are used to measure development and assess their effectiveness
- know and understand the factors, including the globalisation of economic activity, that have led to contemporary differences in development
- describe how and explain why development patterns are changing
- describe and explain how Third World Debt, trade blocs and social constraints and cultural barriers hinder the closing of the development gap
- describe, explain and assess the strategies that exist for reducing the development gap and assess their effectiveness

Summary

Theme 5 Globalisation

1.1 What is globalisation and global shift?

Concepts of cultural, economic, environmental and political globalisation

'Globalisation is the generic term for the process of integration in the realms of trade, economic relations and finance (it is broader, including social relations, knowledge culture and politics) and it is not new. It has been aided by the ICT revolution that has destroyed distance and indeed time. Brands are known the world over and are potentially destroying local diversity.' (Source: Financial Times)

The evolution of globalisation — stages in its development

The ultimate origins of globalisation could be dated back to the Roman Empire, but in the modern sense it evolved via colonialism and the growth of world trade and international financial systems. Its true modern origins lie in the 1975 **OPEC** oil price rises, when the new wealth of oil producers was invested in **MEDC** banks and loaned to developing countries and the emerging industrial economies that had cheaper labour costs.

Globalisation takes four forms:

Economic — the growth of TNCs at the expense of national governments.

Environmental — the creation of global problems that require global solutions.

Cultural — the increase in Western influence, especially American, over aspects such as music and the media.

Political — the increase in influence of Western democracies, the diffusion of state power to regional and international organisations such as the EU and UN, and an increase in the role of non-state actors, e.g. non-governmental organisations (NGOs) such as Save the Children.

The global shift as the movement of economic activities

Global shift involves the physical movement of economic activity from MEDCs, originally to **NICs** (newly industrialised countries), then to **RICs** (recently industrialised countries) and **LEDCs**. Initially the shift involved labour-intensive manufacturing, but increasingly it has involved all sorts of manufacturing and services, especially tourism.

1.2 What factors have led to current economic globalisation?

The factors behind the process of globalisation are outlined below.

Financial factors such as investment

Financial factors contribute, such as **foreign direct investment (FDI)**, where a company has at least a 10% interest in the investment in a receiving country. This investment has been made in order to lock into cheaper production costs (labour, raw materials), and cheaper operating and environmental costs.

Another reason for investment overseas is that companies involved have sought to circumvent import restrictions such as **quotas** and **tariffs** on their goods. One reason why Nissan, a Japanese company, established a factory in Sunderland was to supply the European market with vehicles and thus avoid the payment of import duties into the EU. Several LEDCs have encouraged investment as a way of developing their economies.

Examiner tip

Some questions may focus only on 'the globalisation of economic activity'; others may ask about 'globalisation' in general. The latter type of question will expect you to widen your answer to include the environmental, cultural and political aspects of the globalisation process.

Knowledge check 18

Name one of the first generation of NICs.

Computer technologies

Computer technologies, such as high-speed broadband, the World-Wide Web, videoconferencing and email have speeded up the flow of information and communications. This has enabled business deals to be completed more efficiently and far more quickly.

Transport technologies

The reduction in the price and increase in the speed of transport technology have meant that goods and people can travel further, more cheaply and faster than at any time in history, and with ever-improving comfort and/or convenience. This has reduced the friction of distance and enables companies to locate more economically and take their product to the world market using extremely cheap and efficient transport modes. The tourism industry in particular has benefited from these factors.

The role of the World Trade Organization

The WTO has been working towards promoting free trade between nations and reducing anti-competitive tariffs and quotas that restrict the integration and the flow of goods and services between countries.

Trade blocs

Trade blocs, e.g. the European Union, wield a lot of global power in trading matters. The very existence of trading blocs is a factor that is symptomatic of the process of globalisation.

1.3 How have companies globalised and shifted locations?

Global companies — TNCs/MNCs

The UN defines **TNCs** as corporations that *'possess and control means of production or services outside the country in which they were established'*. Their size is measured in terms of revenues, market capitalisation and, sometimes, employees (Table 2). Most of the world's largest companies are American and include firms in both the manufacturing (e.g. General Electric) and service sectors (e.g. Walmart stores). Significantly, the headquarters of TNCs are concentrated in Brandt's '**North**'.

Table 2 The world's five largest TNCs

World's five largest TNCs, 2010	Turnover in 2010 ($ billion)	Employees in 2009/10
Walmart Stores	408	2,100,000
Royal Dutch Shell	285	101,000
Exxon Mobil	284	83,600
BP	246	80,300
Toyota Group	204	320,000

Knowledge check 19

Give two examples of specific changes in computer technology that have contributed to globalisation.

Knowledge check 20

Give two examples of specific changes in transport technology that have contributed to globalisation.

World Trade Organization An international agency that encourages trade between member nations and administers global trade.

Examiner tip

You should learn details about the structure and organisation of at least one named TNC and use this information to add detail to your examination answers.

The patterns of global manufacturing shift

In 1953, 95% of world manufacturing production was concentrated in the industrialised countries. Although manufacturing remains concentrated in North America, western Europe and Japan, decentralisation has occurred as a result of investment by TNCs in the three generations of NICs in the developing world. NICs have created large companies of their own that are now locating factories in developed countries like the United Kingdom, for example Tata Group, India, who acquired Jaguar Land Rover in 2008.

Location factors for the global shift

Location factors that influence the global shift include the availability of a large, disciplined and skilled workforce, suitable infrastructure, political stability, government incentives and a large domestic market.

Service-sector shifts

Outsourcing or **offshoring** is the global shift of services from MEDCs to NICs, RICs and LEDCs. India has become a common destination for outsourcing due to: Commonwealth links with the UK, high levels of English-language skills, strength of IT education, lower communication costs, and lower wage and capital costs.

The impact of outsourcing and offshoring

Outsourcing and offshoring can bring considerable benefits for countries like India in terms of job creation, higher salaries, greater disposable incomes and a reduction in gender apartheid. However, there are also disadvantages including Westernisation and consequent loss of cultural identity, unsocial hours, and increasing social divisions.

The impact of outsourcing and offshoring for MEDCs is simple: more profitable returns for the participating companies. This enables them to maintain employment in the **quaternary** jobs in their home country and in the manufacturing/service jobs in the production countries. These advantages must be set against significant job losses in the service sector in MEDCs, particularly jobs typically held by women in vulnerable deindustrialised areas.

Deindustrialisation
The decline in the number of people employed in the secondary sector as manufacturing industries have closed. Deindustrialised areas include south Wales, northeast England and the Ruhr in Germany.

1.4 Who wins from the global shift and globalisation?

Global development indicators that identify NICs and RICs

NICs are countries where industrial production has grown sufficiently so that it becomes a major source of national income. Among the first-generation NICs were Hong Kong, Taiwan, South Korea and Singapore. The list of countries that can be called NICs has grown over the past 40 years.

'Recently industrialised country', or RIC, is a term used to cover those countries that have tried to emulate the first-generation NICs. Such countries include Vietnam, Indonesia, Chile, China and India.

GNI per capita in US dollars is the sum of all goods and services produced in a country plus taxes and income from abroad, divided by the population. NICs such as

Singapore and Hong Kong have GNI figures that approach those of MEDCs. GNI was previously known as GNP.

The rise of the NICs/Asian Tigers

Malaysia has developed as a successful second-generation NIC due to:
- Careful state planning, with the government managing the economy and controlling FDI.
- A 'visionary' leader, Dr Mahatir, who led a single-party government for two decades; it had strong controls on the media.
- Taxes from oil in the South China Sea, which provided initial funds for development.
- Low labour costs and laws that favoured investors.
- The growth of export-led industries in economic priority zones (EPZs), free investment zones (FIZs) and free trade zones (FTZs).

Examiner tip
Learn the sequence of development of one named NIC such as Singapore or Malaysia and the physical, economic, social and political factors that have contributed to its success.

Benefits of being an NIC, and benefits to investing countries

Economic benefits for NICs include the expansion of industries and services, increased international trade, rising incomes and infrastructure improvements. Social benefits include an expansion in employment opportunities. Environmental benefits may include the development of ecotourism, aided habitat preservation, and national park development. The benefits to investing countries include efficiency gains and improved profits with lower costs.

The rising superpowers: India and China

Brazil, Russia, India and China together form a group known as 'the **BRICs**' — the fast-growing developing economies. Of these, India and China are seen as emerging superpowers. India's rapid economic growth has been due more to expansion in the service sector than in manufacturing. China's accelerating growth is due mainly to the expansion of manufacturing fuelled by foreign direct investment from Japan, the USA and Europe.

Knowledge check 21
What was the trigger for the rapid growth of manufacturing industry in China in the early 1980s?

1.5 Who loses from the global shift and globalisation?

The negative effects of being an NIC socially and environmentally

- Environmental degradation results from the exploitation of primary resources, e.g. the Carajás iron ore project in Brazil.
- Deforestation has occurred on a large scale. A particular problem is 'the haze', a smog caused by burning forest as it is cleared, e.g. Indonesia.
- Pollution occurs as controls are less stringent: the lack of adequate controls led to the 1984 Bhopal gas-leak disaster in India that caused 22,000 deaths.
- There is a greater divergence in earnings.
- The working conditions of the labour force are often reported to be unhygienic, with very long hours, lack of union representation, no sick pay and none of the social benefits enjoyed in MEDCs.

Knowledge check 22

Give reasons for the outsourcing of ICT requirements of many European and North American firms to Indian companies.

- More women are employed, which can reduce the birth rate to below replacement level.
- Use of expatriate skilled workers, especially in the finance sector (e.g. in Singapore), and in unskilled sectors such as construction can mean that distinct, 'ghettoised' expatriate and immigrant areas develop.
- The population becomes more Westernised, contributing to a lack of cultural identity.

Factors leading to deindustrialisation

Deindustrialisation is the decline in manufacturing that has been experienced in the regions of North America and Europe that industrialised in the nineteenth and early twentieth centuries. This has come about for a number of reasons:

- Many products are at the end of their life cycle — newer products have replaced the old.
- Outmoded production methods have been replaced by newer technologies requiring less labour.
- Labour has clung to old practices and there has been poor labour management.
- There is competition from cheaper locations with low labour costs.
- There was a recession in the 1980s.
- Government support for industries such as coal and steel has been removed.
- Firms have needed to rationalise production.

Tertiarisation The absolute or relative increase of labour in the tertiary sector because of both sectoral change (a shift from manufacturing to services industries) and an increase in service activities within the secondary sector itself.

Changing employment in MEDCs

Responses aimed at reviving regional economies suffering from deindustrialisation may include promoting location, developing leading industries, creating research and development (R&D), providing government assistance at the local, regional and national level, and encouraging tourism. These measures have resulted in the growing tertiarisation of economic activity. Deindustrialised areas of the UK include south Wales and northeast England.

The environmental effects of globalisation

Globalisation has negative effects at global level, such as global warming, and these require global solutions.

Globalisation can also have negative environmental effects at a regional level — as illustrated by the negative effects of being an NIC (above).

Examiner tip

Use your synoptic knowledge of strategies implemented at an international level to address climate change that you covered at AS (Unit G1) to illustrate some of these global solutions.

1.6 What are the causes and effects of political and cultural globalisation?

Empires and superpower status

Globalisation has led to a situation where most countries are interlinked in various ways — politically, culturally and economically. Causes of political globalisation include the influence of the **superpowers**, particularly the USA. The superpowers are often criticised for exploiting the situation of other countries, as in the case of China exerting pressure on the Eurozone for political concessions, such as more representation at the IMF, in exchange for China's financial support.

Westernisation and cultural integration

Cultural integration refers to the increased exchanges of cultural practices between nations. New technologies, such as commercial air travel, satellite television, mass telecommunications and the internet, have created a world where billions now consume identical cultural products, e.g. pop music. People also follow similar cultural practices, such as eating the same foreign food and adopting the same foreign words. The operations of TNCs have resulted in a product and lifestyle monoculture. Effects of cultural globalisation include reduction in cultural diversity, the loss of cultural identity and the development of a **Westernised** consumer culture.

> **Knowledge check 23**
>
> What does the term 'glocalisation' mean?

The rise and re-emergence of other cultures

A rise in **nationalism** and **fundamentalism** has occurred in response to the process of globalisation, with many countries and regions making an attempt to retain their own cultural identities.

Globalisation and the development gap

One of the key negative effects of globalisation has been a growing **development gap**, with increasing disparity in levels of development between countries. This has been fuelled by the uneven pace of development around the world.

Summary

After studying this topic, you should be able to:

- define and understand the terms 'globalisation' and 'global shift'
- know and understand the factors that have led to current economic globalisation, including FDI, changes in computer, communications and transport technology and the role of trade blocs and the WTO
- give examples of the global distributions of named TNCs involved in manufacturing and service

industries, the reasons for these distributions and their impacts

- describe and explain the benefits of globalisation to certain countries, including NICs, RICs, India and China
- describe and explain the disadvantages of globalisation for NICs, MEDCs and the environment
- know and understand the causes and effects of political and cultural globalisation

---

Theme 6(a) Emerging Asia: China

1.1 What are the main physical and demographic characteristics of the country of China?

The People's Republic of China is located in east Asia, on the west side of the Pacific Ocean. It is the third-largest country in the world after Russia and Canada.

A brief overview at the national scale of patterns

(i) Climate

Although most of China lies in the temperate belt, its climatic patterns are complex, ranging from **subtropical** in the south to **sub-arctic** in the north. **Monsoon** winds

dominate the climate and have a major influence on the timing of the rainy season and the amount of rainfall. Alternating seasonal air-mass movements and accompanying winds produce moist summers and dry winters.

(ii) Relief, drainage and water availability

China's relief is both complex and variable, ranging from mostly mountains, high plateaux and deserts in the west, to plains, deltas and hills in the east. The Qinling Mountains provide a natural boundary between north and south China.

The Tibetan Plateau in the west is the source of almost 50% of the major river systems in China, including the three longest rivers: the Yangtze, Huang He (Yellow) and Pearl rivers. These flow west to east, into the Pacific Ocean. About 10% of Chinese river systems drain into the Indian or Arctic oceans. The remaining 40% have no outlet to the sea; they drain through the dry western and northern areas of China, forming deep underground water reserves.

(iii) Natural resources

China has a range of natural mineral resources including coal, iron ore, petroleum, natural gas, mercury, tin, tungsten, antimony, manganese, molybdenum, vanadium, magnetite, aluminium, lead, zinc and uranium. China has the world's largest hydropower potential, with reserves of 680 million kw.

(iv) Population distribution

China's population of 1.3 billion is the largest in the world. It is concentrated in the eastern and coastal part of the country and along major rivers such as the Yangtze, Huang He and Huai. Much of the country, including the steep Himalayas, the dry grasslands in the north, the central region and the Gobi Desert in the north, is almost uninhabited. Nearly 60% of Chinese live in rural areas.

(v) Regional differences in levels of development

Economic growth has taken precedence over equality as the main development goal since the economic reforms of the late 1970s. In the 1980s, under the slogan of 'letting some people and some regions get rich first', China implemented a coastal development strategy. Now, the prosperous urban coastal zone is in sharp contrast to the poor, rural interior.

1.2 Why and how is the economy changing?

Changes in economic policies

After the death of Mao Tse-Tung (Mao Zedong) in 1976, China's economy took a major change in direction. In 1978, Deng Xiaoping, the new leader of the Chinese Communist Party, introduced the 'Open Door' policy, which was designed to overcome China's isolation from the world's economies. The country had become increasingly aware that the world, and south-east Asia in particular, was developing and leaving China behind. China moved towards a **socialist market economy**. Today, China's twenty-first century leaders are focused on economic growth — but on China's terms.

Knowledge check 24

What strategy is in place to transfer water to northern cities such as Beijing?

Knowledge check 25

Why has economic growth in China been concentrated in coastal areas?

The progress made since China's economic reforms were implemented has been significant (Figure 16).

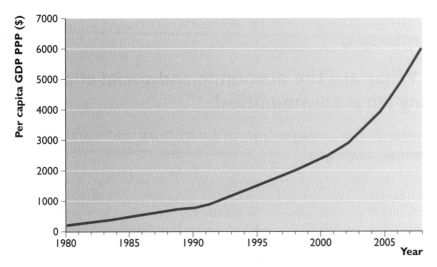

Figure 16 Chinese per capita GDP growth since 1980

New industries in the changing economy

Between 1949 and the late 1970s manufacturing in China was undertaken almost entirely by **state-owned enterprises (SOEs)**. These were mainly heavy industries such as oil, chemicals, power, iron and steel. The 1980s focus on increased productivity forced SOEs towards reform. Large SOEs have improved their management and smaller SOEs eventually privatised. Chinese firms have gradually become more like Western companies.

There has been some success for large SOEs. Industrial output has grown by up to 13% per year and there have been major improvements in terms of technology, managerial skills and efficiency. SOEs have attracted **TNCs** as partners and **FDI** (foreign direct investment) has been highly significant. FDI in China increased from US$3.5 billion in 1990 to US$106 billion in 2010.

Factors affecting the growth of new industries and the contrast between coastal areas and the interior

Since 1979, five special economic zones (SEZs) and 14 **open cities** have been proclaimed. These offer reduced restrictions on land, labour, wages, taxes and planning regulations to overseas firms, especially those involved in high-technology industries. The result has been the emergence and dominance of economic activity in coastal areas — which have received most of the internal investment as well as having imported capital, technology and entrepreneurial skills, at the expense of the interior.

Impact of the changing age structure on the economy

A **positive demographic dividend**, where there are more people of productive age with a low dependency ratio, has coincided with the economic boom. Output per

Examiner tip
Although FDI has been essential for China's economic growth, you should be aware of the recent growth and contribution of Chinese TNCs and domestic industrial developments and markets.

Special economic zones (SEZs) Areas where economic conditions are more favourable for businesses than in other parts of China.

Knowledge check 26
Name one of China's five SEZs.

Knowledge check 27
Name a Chinese TNC.

capita will rise by 10% between 1982 and 2050, but although China's economy is growing it is not becoming more efficient. The **negative demographic dividend**, associated with the rapid process of ageing, will soon affect China. As people live longer they have to either accumulate wealth or face a reduction in their standard of living in their old age.

1.3 What are the economic and social challenges facing rural communities?

Changes in the organisation of agriculture and rural economic activities

The rural economy was the first sector to be reformed because low agricultural productivity posed problems of food security for China. In 1949, all land was brought into communal ownership under the guidance of the state. There was initially an increase in food production as a result of **Green Revolution** farming methods, but this was not sustained. In 1981 farmland was divided up between households, with every household given a 15-year contract to farm the land. This security gave farmers confidence to invest and to manage land more effectively. The reforms revitalised the rural economy. Since that time, agricultural production has increased very slowly and China is beginning to experience food insecurity again.

During Mao's era, rural industries called **town and village enterprises (TVEs)** produced heavy goods such as iron, steel, cement, chemical fertiliser and farm tools, as well as hydroelectric power. After 1978 these enterprises expanded so as to develop a wider range of businesses. Many Chinese farmers preferred to invest their resources in rural industry rather than agriculture. This encouraged the growth of small businesses run by the most successful peasants. Thus a new entrepreneurial class began to emerge and TVEs have become the backbone of development in rural areas.

The effect of population policies in rural areas

The Chinese government's **one-child policy** has been resisted in rural areas, due to the need for labour. The 1980 rural reforms encouraged rural families to demand more family labour and some provinces have conditions for exemption from the one-child policy.

The impacts on and challenges of migration for rural areas

Before the 1980s internal migration was tightly controlled in a bid to avoid unmanageable growth in cities. A household registration system called hukou controlled access to basic requirements such as housing, welfare and employment, and was not transferable between districts. Once one was registered with a rural hukou the registration would be permanent. However, since the 1990s demand in cities for unskilled and semi-skilled labour has led to rural migrants being given a temporary urban hukou, which enables them to have access to some housing and basic welfare in cities. The registration requires annual renewal, so that the authorities can control the number of migrants. Temporary hukou sets the migrant population apart and has increased inequality of access to services in cities.

Examiner tip

It is easy to make generalisations about China, but remember that examiners are looking for detail. Learn some statistics to support your answers and note spatial variations. Although 90% of China's poor live in rural areas, 66% are concentrated in the country's west.

Hukou The Chinese registration system that officially recognises an individual as a resident of an area.

Knowledge check 28

As the majority of the migrants are from the economically active sector, what are the implications of this age-selective migration on rural areas?

Social welfare services such as health and education

There is a significant divide between rural and urban populations, reinforced during the Mao years by the hukou system. Progress and development in urban areas was not matched in rural areas. Consequently, many rural areas are extremely backward, traditional, have very poor services and amenities, and essentially are a world apart from the modern regional and provincial cities. Rural education and health facilities are poor, particularly for an aspiring superpower like China. Villagers often lack any form of social safety-net such as pensions or health insurance.

Sustainable development

In many rural communities the focus on economic growth and new industries is putting pressure on the environment. Deforestation, air and water pollution and the conversion of land from agricultural to industrial use are gradually putting food production in jeopardy. Increased pressure on the remaining farmland increases the risk of soil degradation. Villages and small towns have to increase their own incomes, mainly through small industries, if they are to contribute to health and education services. If these services decline, outmigration will increase and communities will have even greater difficulty in developing businesses and maintaining basic services.

1.4 What are the economic and social challenges facing urban communities?

Changes in the organisation of economic activities in urban areas

Each urban area competes for a share of economic growth and foreign investment. High-technology industrial parks, tariff-free districts and **SEZs** aim to attract new businesses, usually located away from existing, crowded and fully developed urban centres. In some cases, provincial and local governments speculate on development land. The result is thousands of hectares of empty plots of land surrounding one or two high-rise office buildings on the city edges. There is often a lack of coordination between central and local authorities. Some communities are keen to sell their land to developers so that they can change their hukou designation.

Migration to urban areas and increasing social inequality

Millions of people have migrated from rural to urban areas to fill the jobs generated by the economic explosion. Most migrants head for the eastern-seaboard cities of Beijing, Tianjin, Tangshan, Shanghai, Changjiang and the Zhujiang delta. There is also growth in provincial capitals. However, anti-poverty campaigners argue that many workers receive low wages and live in poor conditions. Every year an estimated 200,000 people move to slums on the southern outskirts of the capital, Beijing. Social inequality is increasing between the growing middle class in urban areas and the poor and migrant population. Access to welfare, health and education provision is much better for those residents with permanent urban residential hukou status.

> **Examiner tip**
> Keep up to date with changing Chinese policies. Although urban incomes are now more than three times higher than rural incomes, China's government is taking measures to correct this trend by increasing investment in rural areas, especially in infrastructure, irrigation, education and health.

> **Examiner tip**
> Distinguish between the poverty of rural areas and the poverty of rural migrants who have moved to urban areas, but are denied access to many of the benefits that permanent urban residents receive. 99% of China's poor live in, or come from, rural areas. If migrant workers are excluded from the rural population, 90% of poverty is still rural.

Social welfare services such as health, education and housing

Social challenges associated with China's rapid urbanisation include deprivation and poverty, segregation, and problems associated with health and crime. Housing inequality reflects many of the welfare problems facing urban China.

Increasing rural–urban inequalities

Capital-intensive urban development has created a large productivity gap between the agricultural sector and other sectors of the economy. This has led to a growing gap between rural and urban income per capita. Migration from rural to urban areas, to fill the jobs generated by the economic explosion, has further increased these inequalities. In 2009, China's urban per capita annual income of about US$2,500 was nearly three times that of rural residents. The gap is much more extreme in larger, wealthier cities such as Beijing.

Sustainable development in towns and cities

As towns and cities grow sustainable development becomes more challenging. Demand for energy is rising. People and industries demand more water, so supplies in lakes and groundwater reservoirs fall. As land at the edge of cities is developed for new factories there is less available for farming. City sprawl separates homes from industries, increases the amount of commuting and creates traffic congestion. Domestic and industrial waste disposal facilities come under pressure, particularly from the emerging middle classes. National, provincial and city authorities are not always willing to pay for sustainable projects and services.

Sustainable development
Development that meets the needs of the present, without compromising the ability of future generations to meet their own needs.

Knowledge check 29

Name one of China's planned ecocities.

1.5 What are the effects of globalisation on China?

The role of foreign firms in changing and developing the economy

The liberalisation of trade since the 1980s has led global TNCs to expand aggressively in search of new emerging markets. By encouraging foreign firms into China, competition has raised levels of efficiency and forced large SOEs either to modernise or to close down. **Joint ventures (JVs)** have been vitally important for China, with firms such as Procter & Gamble, Caterpillar and United Technologies being particularly successful. Key features of JVs have been the requirement for technology transfer and an insistence that subcontracted work is given to selected domestic firms. This ensures that China acquires 'know-how' which can then be transferred to domestic firms.

The importance of exports and the role of the WTO

China's export 'basket' consists of labour-intensive export products such as toys, clothes and assembled electronics, as well as more sophisticated products that are more typical of a country with a much higher GDP per capita. The JVs, located in clusters in SEZs, provide a critical source of technology and technology transfer and they dominate exports. China's membership of the WTO since 2001 continues to be a driving force in the opening up of China to both imports and exports. This will have wide-ranging impacts on economic and political systems in China, particularly on the ways in which business is conducted.

The economic and political impacts of China's trade with the rest of the world

China needs resources for its continued economic growth and has been determined to establish trading relationships with countries that can supply raw materials. The increase in food and mineral imports into China has had the effect of driving up many world commodity prices, such as those for iron and other ores. China has increased trade and foreign direct investment on all continents, including into Europe and the USA, and there have been periodic rows with China's trading partners concerned about 'dumping' of exports.

China has always traded with Africa, but recently international attention is being focused on that trade and the political influence that accompanies it.

Within ASEAN (the Association of Southeast Asian Nations) there is some fear about the influx of Chinese goods into local markets. Countries cannot compete with China and therefore must find alternative strategies in order to sustain what economic development they have already achieved.

Knowledge check 30

What are the costs and benefits of Chinese investment in Africa?

1.6 What are the environmental challenges and solutions facing China?

The causes and consequences of soil erosion, industrial pollution, sustainable use of water resources, and the need for energy supplies

It is estimated that since 1949 China has lost one-fifth of its agricultural land to soil erosion and economic development. Air pollution is a major issue in cities such as Beijing and Shanghai due to their heavy reliance on coal. Water shortages are also a problem, particularly in the north, with the result that the Chinese government has embarked on a massive engineering project to transfer water from the wet south to the dry north. Water pollution is a source of health problems due to untreated waste products.

The balance between economic growth and sustainable development

There is growing environmental awareness among grassroots organisations and communities in China, but serious concern for environmental sustainability within the Politburo (the Communist Party's governing body) is still overridden by the desire for economic growth. Despite that, the government response to the Rio and Kyoto agreements on the environment suggested some recognition of the need for sustainability. China signed the Kyoto Protocol in 1998, less than a year after it was set up. This was also intended to establish China as a leader among the developing nations. Environmental concerns are being taken seriously, but bureaucratic problems and some corruption inhibit the implementation of national policies at local level.

Examiner tip

Research the progress made by Chinese manufacturers to develop solar, wind and clean coal technology. This recent trend contradicts the conventional picture of China's poor environmental image.

Summary

After studying this topic, you should be able to:

- provide a brief overview of the main physical and demographic characteristics of China, including regional differences in levels of development
- describe how and explain why the economy of China is changing
- describe and explain the economic and social challenges facing rural communities in China
- describe and explain the economic and social challenges facing urban communities in China
- know and understand the effects of globalisation on China
- know and understand the environmental challenges and solutions facing China

Theme 6(b) Emerging Asia: India

1.1 What are the main physical and demographic characteristics of the country of India?

A brief overview at the national scale of patterns

(i) Climate

It is not easy to generalise about India's climate as the country covers such a large area — India makes up the majority of the Indian subcontinent — and its climate is strongly influenced by both the Himalayas and the Thar Desert. India has six climatic subtypes ranging from desert to alpine tundra. In general, temperatures tend to be cooler in the north, especially between September and March. India has four seasons: winter (January and February), summer (March to May), the wet **monsoon** season (June to September) and the dry monsoon season (October to December).

The wet monsoon season occurs with the movement of the **ITCZ** into the region bringing an area of low pressure and drawing in hot, moist winds from the ocean. Rainfall is increased by **orographic uplift** where these moist winds are drawn over uplands such as the Western Ghats. Temperatures average 30°C and humidity is also very high, with average rainfall around 2000 mm, decreasing with distance inland. Cyclones and hurricanes are frequent towards the end of the rainy season (Figure 14, p. 28).

The cooler dry season coincides with the extension of continental high pressure as the ITCZ moves back towards the Equator and across into the tropics beyond. With high pressure dominating, there is air subsidence and outblowing winds are dry. Temperatures remain relatively high at 25°C in lowland areas and evaporation rates are also high. The weather is much more severe in mountain areas (Figure 15, p. 29).

(ii) Relief, drainage and water availability

The major rivers of India originate in one of three main watersheds: the Himalaya and Karakoram ranges in the north; the Vindhya and Satpura ranges in the centre; and the Western Ghats in the west. The Himalayan river networks are snow-fed and have flow continuously, throughout the year. The other two networks are dependent on the monsoons and have significantly lower discharges during the dry season.

(iii) Natural resources

India's major mineral resources include coal (India has the third-largest reserves in the world), iron ore, manganese, mica, bauxite, titanium ore, natural gas, diamonds, petroleum, limestone and thorium (the world's largest deposits are in Kerala, in the south). Oilfields off Mumbai and onshore in Assam meet 30% of the country's demand; however, India is still heavily dependent on imports of both coal and oil for the rest of its energy needs.

Knowledge check 31

Why is India's dependence on coal as its main energy source unsustainable?

(iv) Population distribution

India's population — 1.15 billion in 2010 — is the fastest-growing in the world and by 2030 India's population will have permanently overtaken China's as the world's largest. The population is concentrated in the fertile northern floodplains. Four states — Bihar, Madhya Pradesh, Rajasthan and Uttar Pradesh — account for 40% of the population and 47% of population growth. The population of Uttar Pradesh, India's most populous state, was 200 million in 2011 — larger than the population of Pakistan. The southern states have lower fertility rates than the north, a contrast which will become more and more marked.

(v) Differences in development between states

India is organised as a federation of states and union territories, each of which has considerable political independence. Since 1991, economic growth in India has been characterised by increased inequality between states — there are high incomes in the northwest (Punjab, Haryana, Delhi, and Himachal Pradesh) and along the west coast (Gujarat, coastal Maharashtra, Goa and Kerala). Lower incomes characterise central and northeast India: eastern Uttar Pradesh, Bihar, Orissa and much of Madhya Pradesh.

Examiner tip

The use of diagrams that are well integrated into your answers and annotated carefully to meet the requirements of the question will always earn you additional credit.

1.2 Why and how is the economy changing?

India has undergone major changes in economic policy. After independence and Partition (whereby British India separated into India and Pakistan) in 1947, India's aim was to develop economically without the participation or influence of foreign capital. Economic policies had a strong anti-export bias. Socialist governments ensured a high level of state control over key industries, which in turn led to excessive bureaucracy and very slow economic growth.

A major economic crisis in 1991 forced the governing Congress Party to borrow money from the **International Monetary Fund (IMF)**. This opened up the economy to economic globalisation. India is now among the ten fastest-growing economies in the world.

Changes in traditional agriculture

Agriculture in India is characterised by unequal productivity across the country. The highest yields are found in Punjab and Haryana and the lowest in the northeastern states of Bihar and Orissa. In the 1960s India was highly dependent on imported food. However, the Green Revolution had a major impact on Indian food production. The area under HYV (high-yielding variety) wheat crops increased from 4 hectares

Green Revolution The application of modern farming techniques, including a package of technology (fertilisers, pesticides, water control and mechanisation), to developing countries.

in 1963 to 4 million hectares in 1971. The wheat revolution was followed by similar changes involving rice, sugar, millet and oilseed crops, as well as cotton. As a result, food production exceeded population growth. The Green Revolution was successful in raising incomes for farmers on naturally fertile soil, but increased inequalities between wealthy farmers on productive land and poorer farmers on marginal land.

The role of agribusiness

Agribusinesses play an increasingly significant role in agricultural exports and in food security. They control much of the chain, from seeds and fertilisers to finance, distribution and marketing. The increased output from new, large farms in India helps to maintain food security.

The growth of service and financial industries

An extensive financial and banking sector supports the rapidly expanding Indian economy. India has a wide and sophisticated banking network. The sector includes a number of national and state-level financial institutions and a well established stock market. The Indian capital markets are rapidly moving towards a modern market including derivative and internet-based trading. The services sector, including financial services, software services, accounting services and entertainment industries like Bollywood, contributes 41% to GDP.

Factors affecting the growth of manufacturing industries

In addition to political changes, factors responsible for the rapid growth of manufacturing industries include economic change, such as the emergence and investment policies of **TNCs**. Other contributing factors are the growth in Indian firms and of an urban, educated, middle-class population whose members have become consumers themselves and who provide a large market for new consumer goods. Technological factors have also played a significant role, particularly the speed and distance over which communications and movement can now take place due to changes in computer, transport and communication technologies.

The need for major developments in infrastructure throughout India

India has serious transport issues. The road transport sector has been declared a priority and will have access to loans at favourable conditions. The country needs US$200 billion worth of new ports, roads and other infrastructure and another US$50 billion to modernise 40,000 km of roads. Currently there are significant delays in distribution and severe bottlenecks, with state and federal governments often in opposition. The National Highways Act has been modified to help reduce tolls on national motorways, bridges and tunnels. The government is also implementing a new policy that aims to improve India's telecommunication systems.

Knowledge check 32

Why is India's poor infrastructure regarded as a key obstacle to its growth?

Knowledge check 33

In what ways has India's growth been different from that of other Asian NICs?

1.3 What are the economic and social challenges facing rural communities?

The traditional socioeconomic characteristics of rural India

'India lives in its villages' is a quote from the Registrar-General (2005). Although our impression of India is of overcrowded cities, in 2001 the average Indian lived in a settlement of 4,200 people. Some 72% of the population is classed as rural, with 58% being farmers. While in some ways India is rapidly becoming a middle-class country with Western lifestyles, in other ways it remains a rural country where social and religious traditions are embedded.

Most rural Indians have lower educational levels, higher mortality and fertility, greater poverty, and access to fewer services and amenities than urban dwellers. Most Indians live their whole lives in a relatively limited geographical area. Some rural areas in the states of Bihar, Jharkhand, Uttar Pradesh and Orissa are officially destitute.

The impacts of migration

Migration to cities has occurred due to a lack of opportunities in rural areas. This push factor affects the sustainability of urban growth, which has become a concern for state and national governments. Development of rural areas could do much to stem internal migration and take pressure from urban centres.

Social welfare services in rural areas

There are many welfare concerns, such as the need to provide minimal social and income security for agricultural workers. Education is a challenge in many rural areas, particularly the education of girls, and dropout rates are high and attendance is poor. Poorer agricultural households show the worst attendance levels, especially in migration and harvest seasons.

Food production and combating hunger after the Green Revolution

Economic challenges associated with India's rural communities include the challenges of food production and land reform and problems with infrastructure and service provision. **Social challenges** include population growth, hunger, malnutrition and migration.

The relationship between population growth, hunger and sustainability in rural India

India's growing population increases the challenge of food security. There is growing demand for fertile farmland to be used by multinational companies to grow industrial and food crops for export. New industries also demand land in order to expand. Both of these developments also use up water resources and increase pollution of soil and water.

Examiner tip
It is easy to make generalisations about India, but remember that examiners are looking for detail. Learn some statistics to support your answers and note spatial variations.

Examiner tip
Although states such as Kerala, Tamil Nadu and West Bengal have made progress in reducing levels of rural poverty, Assam, Bihar, Madhya Pradesh, Maharashtra and Rajasthan are still among the eight poorest.

Poor farmers are often forced onto more marginal land that, without expensive fertilisers and pesticides, produces lower yields. Those farmers who do try new technologies are at risk of debt if crops should fail. With more people to feed and less quality land available, the poor face an increased risk of hunger.

A key question for India is how to provide sufficient food for an expanding population while at the same time encouraging agricultural and industrial development. Individual states have very different attitudes to this problem.

1.4 What are the economic and social challenges facing urban communities?

The rapidly growing and economically powerful urban areas contrast with traditional and often remote villages. Only 11% of India's population live in cities of over 1 million people. The number of city dwellers in India increased by 60 million between 1991 and 2001, but the number of rural dwellers increased by 113 million. There is a crisis in urban infrastructure due to that extra 60 million, but also due to the aspirations on the part of the middle classes for private cars and a higher-consumption lifestyle.

Changes in type of economic activity

Global Mumbai is classified as a global city (also called world city or sometimes alpha city), which is a city that is deemed to be an important node in the global economic system.

Indian cities with over 1 million people are integrated into the global economy but smaller cities tend to look towards the local economy.

Migration to urban areas and the interdependence of rural and urban populations

In India, rural–urban links are strong. Journeys of 20–30 km into Mumbai, Kolkata and Chennai are not uncommon. Family members return to rural homes to help with the harvest, and money sent by urban dwellers to rural areas makes a significant contribution to poverty reduction despite increasing urban poverty. The reverse also happens, with rural families sending money to urban relatives in order to support students or men seeking work.

Delivering modern infrastructure and social welfare services

Examiner tip
In any discussion of intra-urban inequalities, provide place specific detail and avoid restricting your answer to only one city. In Mumbai 60% of the population live in slums on 6% of the city's land. In Delhi one-third of the 17 million residents do not have access to city water.

Social challenges include deprivation and poverty, segregation, and problems associated with housing, health and crime. As a result, most urban settlements are characterised by shortfalls in housing, inadequate sewage, poverty, and social unrest. This makes urban governance a difficult task.

Increasing inequalities within urban areas; the informal sector and urban poverty

The poor compete with middle-class people for land, which may be wanted for building homes, transport installations and retail malls. Economic challenges associated with India's urbanisation include the growth of the informal sector, problems of service provision, and exploitation of the labour force.

Sustainable development in cities

Five Indian cities are included in the global Sustainable Cities Programme: Chennai, Bangalore, Hyderabad, Delhi and Kolkata. The key development issues for the sustainable development of cities are to secure housing rights, to provide access to civic amenities, public health, and education, to provide safe and secure drinking water, and to improve food security. In addition, there needs to be freedom from violence and intimidation and the provision of adequate social-security programmes.

1.5 What are the effects of globalisation on India?

The impacts of global trade on the national economy

Since 1991, the economy has undergone a major transformation. The high levels of protectionism have been replaced by growth in exports. India is building economic and political ties around the world. It has considerable influence over world trade as a founder signatory of the General Agreement on Tariffs and Trade (GATT), the forerunner of the WTO. India leads the developing nations in global trade negotiations and is trying to encourage a more liberal global trade regime, especially in terms of services. India is one of the top ten exporters of services in the world and is famed for its specialist trade in IT services. The direction of trade is also changing, away from Russia and eastern Europe towards the USA, EU and east Asia. India's major trading partners are the USA and China, but it is also developing trade links with African countries.

The effects of the 1991 debt crisis and structural adjustment

As noted, a major economic crisis in 1991 forced the governing Congress Party to borrow money from the International Monetary Fund. This triggered a major change in the economy, allowing direct foreign investments into the country, which opened India up to economic globalisation.

The growth of Indian TNCs

Some industries, such as defence and aerospace, remain under state control, but many manufacturing sectors, including vehicle, consumer electronics and white-goods manufacturing, are now open to foreign direct investment. Indian companies may set up joint ventures or become wholly-owned subsidiaries of foreign firms.

Conflicting views of the benefits of globalisation for India

Beneficial effects of globalisation on India include foreign investments into pharmaceutical, petroleum and manufacturing industries, which have provided a significant boost to the Indian economy along with new employment opportunities, and have contributed to the reduction in levels of unemployment and poverty. Foreign companies bring advanced technology, helping to make Indian industry more technologically advanced. As a result India has seen an increase in international trade with a growth in exports, rising incomes, and infrastructure improvements.

Examiner tip

When answering questions about sustainable development in cities avoid discussing urban problems in general and ensure that you look at the extent to which the social, economic and environmental conditions in urban communities meet the needs of the present and future generations, and provide precise locational information.

Knowledge check 34

Give an example of an Indian TNC.

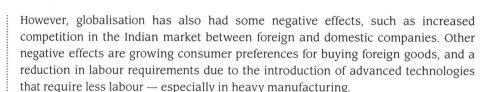

However, globalisation has also had some negative effects, such as increased competition in the Indian market between foreign and domestic companies. Other negative effects are growing consumer preferences for buying foreign goods, and a reduction in labour requirements due to the introduction of advanced technologies that require less labour — especially in heavy manufacturing.

The impact of globalisation on India's poor

Globalisation has led to widening social and regional disparities. India's economic development, particularly since 1991, has increased inequality between castes and between states. Although India has developed in terms of rapid industrialisation, consumerism and materialism, in many cases this 'progress' has not affected everyone. In fact, many people are worse off than in 1991 — especially the *Adivasi* (the scheduled tribes and castes and indigenous communities), the landless peasants, and marginalised farmers. The real challenge for India is how to enable the benefits of economic growth to trickle down to the very poor. If India continues to grow at 8% per year, average rates of poverty will fall to single figures within 20 years. This would be a significant achievement but would still mask many underlying economic inequalities across the nation.

1.6 What are the environmental challenges and solutions facing India?

The causes and consequences of deforestation, soil erosion, industrial pollution in major cities, sustainable use of water resources, and the need for energy supplies

After independence India's forests were exploited commercially for pulp and paper, and they continue to be overexploited. Huge areas have been deforested, accompanied by serious soil erosion. There has been significant replanting by state forestry departments, but this has been primarily of commercial timber — eucalyptus, teak and pine. Although the forested area is increasing, the loss of biodiversity is not sustainable.

The Ganga (Ganges) River, which has great spiritual and emotional significance for Indians, is seriously polluted. In 1984, the Department for the Environment set up an action plan to reduce pollution. Initiated in 1986, the plan aimed to intercept, divert and treat 882 million litres of waste water per day. Recognising the need to revamp the river conservation strategy, the central government has given Ganga the status of a 'National River' and in 2009 it set up the National Ganga River Basin Authority (NGRBA). This is an empowered planning, financing, monitoring and coordinating authority for effective abatement of pollution and conservation of the river.

India has both too much and too little water. The areas that flood in the monsoons, and the number of people that are affected, increase every year. Increased demand in rural areas results from the demands of agriculture, rising standards of living and rural industrialisation. The unpredictability of rainfall in many semi-arid areas is increasing because of failed monsoons and climate change (covered at AS, Unit G1).

India faces a critical challenge to meet a rapidly increasing demand for energy. India is sixth in the world in terms of energy demand. There are significant reserves of coal, but limited oil and gas resources. However, India has considerable potential for exploitation of renewable energy resources such as solar and wind power, and biofuels (from sugarcane).

The balance between economic growth and sustainable development

The challenges of addressing poverty as well as managing the environment sustainably remain significant. After the UN Conference on Human Environment in 1972, India created a National Committee on Environmental Planning and Coordination (NCEPC). Environmental issues were then included in the national 5-year plans. In the 1980s the Ministry of Environment and Forests was created. Now, numerous autonomous agencies, offices and institutions exist, set up by the government at national and state level.

There is a will to improve the environment, but this often conflicts with other demands and, as in most other countries, sets government departments at odds with each other.

India faces many of the same environmental issues as developing countries. It is challenged by the need to meet the demands of industrialisation for development while understanding the necessity for environmental sustainability. The rates of urban and rural change make it hard to ensure that the best environmental decisions are taken. India's democracy can hinder progress. The sheer scale of the environmental challenges is daunting, yet progress is being made at national and grassroots levels.

Summary

After studying this topic, you should be able to:

- provide a brief overview of the main physical and demographic characteristics of India, including differences in levels of development between states
- describe how and explain why the economy of India is changing
- describe and explain the economic and social challenges facing rural communities in India
- describe and explain the economic and social challenges facing urban communities in India
- know and understand the effects of globalisation on India
- know and understand the environmental challenges and solutions facing India

Section B: Individual research enquiry

Preparing for Section B

The structure of the individual research enquiry

Table 3 gives an indication of the timetable you should be working to. Much will depend on whether your school or college has entered you for a January or a June examination.

Table 3 Individual research enquiry timetable

January entry	Timetable	June entry	Timetable
Teacher will introduce the research themes for year	June–July at end of AS	Teacher will introduce the research themes for the year	June–July at end of AS
Confirm what topic you have selected	July before term ends	Confirm what topic you have selected	September or December
Do the research	August and/or September Year 13	Do the research	October–December Year 13 or January–February Year 13
Data processing, structuring report and writing up	October–November Year 13	Data processing, structuring report and writing up	February–April Year 13
Have your work internally assessed to make sure that it gives you a chance to answer questions about it	December	Have your work internally assessed to make sure that it gives you a chance to answer questions about it	April–May
Revise your research to enable you to answer questions	First weeks in January	Revise your research to enable you to answer questions	Late May
Sit examination	Late January	Sit examination	Early June

Pre-decision research

This is a key stage in your work. In deciding what topic area you wish to research, ask yourself the following questions:

- *What are my interests and what other A2 subjects link to the list of research themes?* For instance, if you are studying psychology, Environmental Psychology might overlap with some of your studies. Alternatively, the Geography of Crime, and Deprivation, may interest sociology students. Chemistry students might be attracted by Atmospheric and Water Pollution.
- *Which theme interests me?* If your class size is small, expect there to be restrictions on your choices.
- *Have I discussed my preferred theme with the teacher and can the teacher support it?* Some departments will restrict the choice of themes for logistical reasons.
- *Do I have a potential topic of research on the selected theme?* Some suggestions are available on the WJEC website. These are updated each year (**www.wjec.co.uk/index.php?subject=56&level=21**). Table 4 provides some examples.
- *Have I found or been given background reading on the theme?* This should include articles and textbooks, as well as internet searches.
- *Are there others in the class researching similar topics?* If there are, be prepared for the teacher to expect you to carry out some shared group background research and data collection tasks.
- *Have I set myself a timetable for my research?* Things cannot be left until the last minute. The timescale is tight. Remember that research is 90% perspiration and 10% inspiration!
- Finally, *do I understand the route/sequence of enquiry?*

The route/sequence of enquiry

There are five stages that make up the framework for the investigation:

(1) Planning stages of the investigation.

(2) Data Collection.

(3) Data Refinement and Display.

(4) Description, Analysis and Interpretation.

(5) Conclusion and Evaluation.

These are elaborated in greater detail in the WJEC publication, 'Enquiry approach which can be applied to G1, G2 and G3B' (**www.wjec.co.uk/uploads/publications/9232. doc**).

Examination questions can focus on any aspect of this process or ask about a theme in your research, such as the use of ICT or numeracy.

Planning, background research and topic title setting

All good background research involves reading around the subject and solid preparatory work.

- One starting point could be undertaking internet research using the topic area title. In using the web, make sure that you know where the information came from and how valid the source is, as well as how biased it may be. Companies and organisations with reputations to maintain might structure information so as to be seen in a favourable light. The Royal Geographical Society website has a link to *Geography Now* (**www.rgs.org/now**), which has articles on topics such as 'putting poverty in its place', an ideal background for work on deprivation, or 'restoring urban rivers', potentially providing a good idea for a researcher on Rivers. Urban-based research on topics to do with deprivation, retailing, environmental psychology and crime could make use of Google Streetview, which has video footage of streets in most urban areas and many smaller settlements in the UK.

- Read a textbook on the topic. For instance, *Managing Ecosystems* (A. Kidd, Hodder & Stoughton, 1999) could be a start for work on small-scale ecosystems. The Geographical Association's *Discovering Cities* series provides ideas on the themes of deprivation, retailing and leisure in cities such as Bristol, Liverpool, Manchester, Portsmouth, Sheffield and Nottingham.

- Consult the journals in your institution's geography department. The following are useful sources of information:
  - *Geography Review*. For instance, the May 2004 issue had an article titled 'Investigating coastal sand dunes' — ideal for the Small-Scale Ecosystems theme.
  - *Topic Eye*. Articles in the 2012–13 series focus on malaria, regional variations in health and tourism as a development process.
  - *The Geographical*. For example, the February 2009 edition had an article by P. Thomas, editor of *The Ecologist*, titled 'The trouble with travel' — potential inspiration for Leisure and Recreation research.

- If your study involves fieldwork, what equipment do you need? Can the fieldwork be completed by you on your own or would you need assistance from friends and classmates? Do you know how to use the equipment? Does the school permit you to use the equipment without staff supervision?

- If the topic requires fieldwork, select and visit the site(s) for your study.

- Prepare any datasheets that you may need to record findings. If you are going to use questionnaires, prepare the questions and pilot them on your teachers and

> **Examiner tip**
> Make sure that you can show that you planned the enquiry carefully.

colleagues. What type of sampling will you use? For the examination you will need to know the merits and drawbacks of the different types of sampling.

- What are the risks involved? You should assess these and your school should help you with this. If it is a topic involving Rivers or Small-Scale Ecosystems you may need to check the weather forecast; if studying tidal rivers or coastal salt marshes, check the tide times.
- Decide when you are going to do the fieldwork or when you are going to spend time on your research.
- Always ensure you have permission to enter any property or area.
- Make the topic straightforward and not too long or complicated. It must be manageable. Table 4 illustrates some straightforward topics.
- Make sure that you allocate sufficient time for:
  - data collection
  - revisiting sites or sources to improve rigour
  - writing up your research
- Create a 'time budget' with your teacher for the work.

Table 4 Potential topics for your individual research enquiry

Potential topic	Research theme
Is there a relationship between the amount and quality of green space and residential preference?	Environmental Psychology
Is there a relationship between building density and temperature in an urban area?	Microclimates
Does retailing provision vary according to the size of rural settlement?	Geography of Retailing
How important is the role of water in the transmission of malaria?	Geography of Disease
Patterns of drink-related crime and policing	Geography of Crime
Variations in river water quality	Atmospheric and Water Pollution
Contrasts in shoreline vegetation patterns	Small-Scale Ecosystems

Of these topics, several would involve fieldwork and primary data collection, while the fourth topic mentioned is dependent on good secondary sources.

Collecting the information: data collection methods

Do you know from where you are collecting the information? Do you know why you are collecting the information? What part of the topic is it intended for? Can you describe your methods of gaining information and data, and justify your use of these methods?

Presenting the information and writing the report

Once you have your data you will need to process and interpret the findings. If several of your colleagues are doing related topics, you could meet up, either with or without the tutor, to exchange information. Such a 'seminar' format will enable everybody to gain the maximum amount of understanding from the research and will probably provide you with ideas that you had not previously considered. Expect the tutor to guide you but not to provide you with all the answers.

It often pays dividends to graph, map or statistically process your information. An annotated location map for the study is essential because in the examination you will have to say what you researched and where you did the research. This is especially important in the case of primary-data and field-based studies. The Geographical Association has two relevant texts that you should consult: *Methods of Analysis of Fieldwork Data* and *Methods of Presenting Fieldwork Data*.

What kinds of diagrams should you consider?

- **Graphs**. Line graphs, histograms, pie charts (but do not overdo these), scatter graphs, frequency curves, long sections, cross-sections, and triangular graphs.
- **Maps**. Locational, choropleth, isoline, flow, distribution, and relevant published maps, such as Goad plans for shopping centres.
- **Photographs**. These can include a wide range of images, including satellite imagery, oblique and vertical air photos, Google Earth and Google Streetview images, or your own photographs. The Panoramio online photo-sharing community (**www.panoramio.com**) provides another source of good images.
- **Statistical techniques**. Statistical and test data may be appropriate for your study. Mean, median, mode, standard deviation, scaling and weighting of data, conflict matrices, Spearman Rank, chi-squared and interpreted bipolar analyses are all useful. You should understand the sampling methods that you used to gain data.

Interpreting the information

Once the data have been satisfactorily collected, draw tentative conclusions from the data and assess the validity of the conclusions. This evaluative phase is essential because questions will be asked on this aspect of the report.

Writing up the report

A written report is not compulsory but it has the benefit of drawing together all the elements of your research investigation. By writing up a report of approximately 2,000–2,500 words, you will have produced both a completed study and, most importantly, a revision document. The report will provide the source of knowledge, understanding, application and skills used by the examiner to award you a mark (Table 5).

Examiner tip

Knowing about your study is very important. Analysing, interpreting and evaluating geographical information, issues and viewpoints, and the ability to relate and apply your study to the broader field of geography, will gain you credit.

Table 5 How the marks are allocated for both parts of Section B

	Knowledge and understanding	Application	Skills	Total
Unit G3 Section B Total mark	9	6	10	25
Part (a)	4	2	4	10
Part (b)	5	4	6	15

Preparing for the examination

Let your teacher assess your work because he or she will be able to guide your revision on the topic towards the examination. The potential questions on the topic area will relate to both the methodology of the research study (part a) and the findings of your own research (part b). The examination assumes that you have completed individual

research/investigative work, which may have included fieldwork. It assumes that you can identify and analyse the connections between the different aspects of geography, and analyse and synthesise geographical information in a variety of forms and from a range of sources. Crucially, the examiner expects you to be able to critically reflect on and evaluate the strengths and limitations of your enquiry at all stages: in the planning, data collection, data presentation, results, conclusions, and final topic re-evaluation stages.

Knowledge and understanding

To gain marks for knowledge and understanding, which comprise 50% of the marks in part (a) and 46% of those in part (b), you should be able to demonstrate in your answers that you:

- have an understanding of the key concepts that apply to your theme
- know the meaning of terms that are relevant to your study
- understand the processes that are operating in your study
- understand the interrelationship between factors in your study
- can explain where you undertook your study, how long it took and what conclusions you reached
- understand the techniques of data collection that you used and the techniques that you used to interpret the information
- know about other studies within your topic area

Application

You will gain credit if you can show in your answers that you know how to analyse, interpret and, above all, evaluate the information that you have researched. Critical reflection on and evaluation of the potential of your approach to the topic, and the limitations of your approach and methods, will demonstrate your ability to apply yourself to the topic. When researching the human themes you will also have the opportunity to gain credit for understanding people's viewpoints.

Skills

The final section of the marks will focus on the skills that you demonstrate in your answers — primarily the skill of carrying out research. Communicating your research findings in an examination requires the skill of writing in prose. Questions may also test your understanding of the data collection and processing skills used during your investigation.

The G3B examination

All the questions in Section B will examine the same topics no matter which research theme you studied.

The questions in **part (a)** will focus on aspects of the sequence or route of enquiry. You may be asked about aspects of data collection; for example: *Describe and justify different methods of acquiring information that could be used to investigate the distribution of crime* (an essentially identical title will appear for every research theme — no matter whether it is, for example, Geography of Disease, or Rivers). Other questions on data collection could focus on how you avoided bias, methods of sampling, or a critical evaluation of data collection techniques. It is essential that you state the title of your investigation at the start of your answer to **part (b)** of the question. In this part you will always be asked to write about the findings of your individual study.

Here are some questions that could be set using *The route or sequence of enquiry* stages outlined on pages 62–63.

- How did you plan your personal research investigation?
- Critically evaluate the planning that you undertook for your personal research investigation.
- Evaluate the primary or secondary sources that you used in your personal research into...
- Describe and explain the sources that you used in your research investigation.
- How did you avoid bias in either your own primary data collection or the secondary sources that you used?
- How did you organise the data collected to answer your topic's aims?
- Evaluate the methods of presenting data that you used in your personal research investigation into...
- Explain how you used the data that you collected in your personal research to investigate your chosen topic.
- Critically evaluate the data collection used in your personal research.
- What were the conclusions that you drew from your research investigation?
- Evaluate the conclusions that you drew from your research investigation. How certain can you be that they were valid?

You could ask yourself some of the following questions about your enquiry.

- Did the conclusions of your research investigation match those in the literature that you read before you started? If not, why not?
- Have I read and researched the theme and topic area thoroughly?
- Have I planned the enquiry carefully?
- What sources of information do I need?
- When, where and how will I collect my data?
- Have I presented the data using a variety of appropriate techniques?
- What do the data show me?
- What are the main findings of my research?
- What have I learnt from undertaking my research?
- What are the limitations of my enquiry?
- How could my enquiry be improved?

Examiner tip
Many of these questions are evaluative, i.e. you are expected to discuss both the good and less-effective aspects of your work.

Remember

The individual research enquiry is intended to be an exercise in which you *learn for yourself with teacher guidance*. It is based on the route or sequence of enquiry (see pages 62–63). Whatever you do during the research investigation, always maintain a critical appreciation of data and any potential bias. The investigation is small-scale so do not become overambitious. Consult your teacher on a regular basis and seek guidance. Do not do the work by merely copying from the internet — this will not help you to answer the examination questions. Remember also that your teacher will not know exactly what questions will be set in the examination, so do not rely on his/her predictions for your own revision.

Questions & Answers

Preparing for the unit test

G3A examines your knowledge and understanding of two contemporary themes you have studied and will test your ability to identify and analyse the connections between these themes and the relevant aspects of geography already studied as part of your Geography course — known as **synopticity**. **G3B** examines your ability to carry out research into a selected topic where the synoptic links to the themes of this paper and those at AS are important. In order to achieve an A* at A2, you need to demonstrate to the examiner that your knowledge of the contemporary theme you have studied is very sound and that you have a very firm grasp of the definitions and concepts that are central to the topics you have covered.

In addition, you need to acquire a sound knowledge of supporting, relevant **case studies** and the ability, where appropriate, to produce clear, relevant and accurate annotated diagrams that are effectively integrated into your answers. Using the correct vocabulary is also essential. It may be useful to compile a list of **key terms**: use textbooks, teacher's notes and your own class notes to build up a glossary of relevant geographical terms. The use of the examiner tips, knowledge checks, definitions and summaries in this book will also help to improve your subject knowledge and understanding, as well as examination technique.

As this is an A2 unit you will be expected to read around the contemporary theme that you are studying to give you a broader and better-informed knowledge base. Look for relevant articles in subject-specific magazines such as *Geography Review* and *The Geographical*, as well as relevant geography textbooks, quality newspapers and websites, such as NGfL (**www.ngfl-cymru.org.uk/eng/vtc-home/vtc-aas-home/vtc-as_a-geography**).

The unit test

Timing

The examination lasts for a total of 2 hours 15 minutes and counts for 60% of your A2 award, but it is broken down into two parts:
- Section A: Contemporary Themes in Geography, lasts 1 hour 30 minutes.
- Section B: Research in Geography, lasts 45 minutes.

The two parts of the examination will be handed out separately in the examination room. *Section A answers will be collected in by the exam invigilators before you are given the separate paper for Section B.*

In Section A each question is worth 25 marks and you are required to answer two questions and divide up your time equally between them, giving you 45 minutes to answer each question. In Section B you will need to answer a compulsory two-part question from the research theme and topic area that you have selected. The part (a) question is a 10-mark generic question that examines your understanding of the enquiry approach, whereas the part (b) question is a 15-mark question that will examine the findings of your own particular research topic. You will therefore need to spend 18 minutes (just under 20 minutes) on question (a) and 27 minutes (just over 25 minutes) on question (b).

Choice of questions

Section A

Section A is an essay paper, which is divided into two parts, each made up of eight questions. You must answer one question from Contemporary Physical Themes 1–3, and one question from Contemporary Human Themes 4–6. Each question is worth 25 marks or 40 UMS (total 80 UMS).

Themes 1–3 (Questions 1–8) are set on the following Contemporary Physical Themes:
- Theme 1 Extreme environments: deserts and tundra
- Theme 2(a) Glacial landforms and their management *or*
- Theme 2(b) Coastal landforms and their management
- Theme 3 Climatic hazards

Themes 4–6 (Questions 9–16) are set on the following Contemporary Human Themes:
- Theme 4 Development
- Theme 5 Globalisation
- Theme 6(a) Emerging Asia: China *or*
- Theme 6(b) Emerging Asia: India

Section B

This part of the specification provides you with the opportunity to carry out individual research and out-of-classroom work, including fieldwork, on a pre-set topic on one research investigation theme listed below.

You are required to answer a compulsory two-part question from the research theme and topic area that you have selected. It is possible that you will have carried out research on a different topic from others in your class/group as you will have been encouraged to research a topic that particularly interests you. However, your tutor may have given all of your class the same research topic. This compulsory two-part question is worth 25 marks or 40 UMS.

This section of the examination is your opportunity to carry out work in an area that interests you personally. It is work that is inspired and guided by your teacher, but allows you to demonstrate that you can research a particular topic either in the field, or through desk-based research in the library and on the internet.

You have the opportunity to opt for one of ten themes listed in Table 6, each of which will have a topic area for the year when you take your A2 examinations (whether the exam is in January or June, the topic area will be the same). You have to research only *one* topic area. You will be then examined on the topic area that you researched.

Table 6 Themes and exemplar topic areas

Theme	Topic area 2012	Topic area 2013
Geography of Crime	The management of crime	Crime in rural areas
Deprivation	Deprivation in urban areas	Spatial and/or temporal changes in deprivation
Geography of Disease	The impacts of a human disease	Spatial variations in disease
Environmental Psychology	Gender and environmental perception	Impact of the environment on behaviour
Leisure and Recreation	Use of green spaces	Rural/urban contrasts in leisure and recreation
Microclimates	Woodland microclimates	Upland microclimates
Atmospheric and Water Pollution	Air pollution	Pollution of the marine environment
Geography of Retailing	Clone towns	Out-of-town shopping
Rivers	Flooding	Landforms in river valleys
Small-Scale Ecosystems	Woodland ecosystems	Hydroseres and/or haloseres

You have to select one topic area and carry out your research under a title within the area that you and your teacher agree is suitable. You may wish to submit your chosen title for approval by the Principal Examiner by completing a proposal form on the WJEC Geography website (**www.wjec.co.uk/index.php?subject=56&level=21**). It is important that your title is carefully selected in order to produce research of a high quality because the examination question carries 40 UMS, or 10% of your total A-level marks.

How answers are marked

Section A questions are marked on five levels.

Level	Description of quality	Marks range	What examiners are looking for
5	Very good	21–25	Very good knowledge and understanding used critically. An ability to evaluate arguments. Good, possibly original examples. A clear, coherent essay which is grammatically correct. Good diagrams and sketch maps where appropriate.
4	Good	16–20	Good knowledge and understanding with some critical awareness. Evaluation more patchy. A clearly structured essay that uses good English but argues points soundly rather than strongly. Appropriate diagrams and maps not always fully labelled.
3	Average	11–15	Knowledge and understanding present but some points may be partial and lack exemplar support. Mainly uses text or taught examples of variable quality. Language is straightforward and is in need of more complex use of geographical arguments.
2	Marginal	6–10	Some knowledge and understanding but with gaps and misconceptions. The answer is not very broad in its cover and has only limited support from examples, diagrams and maps. Language is variable and slips occur.
1	Weak	1–5	Knowledge and understanding is very basic and lacking supporting evidence. Shows evidence of not understanding the question. Use of language of geography and written style contains slips.

A fuller, official copy of the generic mark schemes can be found in the Specimen Assessment Materials at your college or school.

Section B contains 10 two-part questions marked out of 10 (part a) and 15 (part b).

Part (a) is marked on three levels.

Part (a)	What examiners are looking for
Level 3 **8–10 marks**	Very good knowledge and understanding used critically which is applied to research route to enquiry. The work is obviously based on research and uses it to provide good supporting evidence. A clear, coherent mini-essay which is grammatically correct.
Level 2 **4–7 marks**	Good knowledge and understanding with some critical awareness of the route to enquiry. A clearly structured mini-essay that uses good English but argues points soundly rather than strongly. Appropriate diagrams and maps not always fully labelled.
Level 1 **1–3 marks**	Some limited knowledge and understanding of some aspects of the route to enquiry present but some points may be partial and lack exemplar support from the research. May use taught material of variable relevance. 'All I know' rather than an answer to the question. Language is variable, lacking paragraphs and may have weak grammar and syntax.

Part (b) is marked on four levels. The levels will partially relate to the expected content in your answers.

Part (b)	Characteristics of level (it is not necessary to meet all of the characteristics to be placed in a level)
Level 4 **13–15 marks**	Provides title of research. Very good knowledge of the topic studied and a critical awareness of the route to enquiry as applied to the topic in question. Provides very good support from own research. May have some good diagrammatic material and maps to support answer. Written in a sound, coherent essay style, which is grammatically correct, with a sequence of ideas that enables the question to be answered fully. Able to evaluate when required. Concludes in relation to the question.
Level 3 **9–12 marks**	Provides title of research. Good knowledge of the topic with some gaps. Understanding of the route to enquiry is present but may occasionally be unsound. Good knowledge and understanding with some critical awareness. Evaluation more patchy. A clearly structured essay that uses good English but argues points soundly rather than strongly. May have a conclusion.
Level 2 **5–8 marks**	Provides title of research. Knowledge and understanding present but some points may be partial and lack exemplar support from research theme studied. Verges on the formulaic answer which is possibly almost identical for all students in centre. Language is straightforward and will possibly lack paragraphing. Perhaps going off at a tangent with an 'all I know' answer. Tails off and may not have a conclusion.
Level 1 **1–4 marks**	Possibly neglects to state title. Some knowledge and understanding but with gaps and misconceptions that indicate an inability to understand the question. Evidence that the research was superficial. Only limited support from research. Language is variable and slips occur.

Quality of written communication

In addition to assessing your subject knowledge, examiners are required to give credit for the quality of written communication. In this respect examiners are looking for:

- the ability to write clearly and grammatically
- the use of correct spelling
- clear paragraph construction
- appropriate use of geographical terminology
- coherent structuring of the answer with paragraphs linked logically so that the line of discussion being taken is clear to the reader
- reference in the text to supporting diagrams so that they become an integral part of the total response

Examination skills

It is important at the outset to work out very clearly the requirements of the question you have decided to answer. It is rare that learnt material can simply be reproduced.

It is important that you select only those parts of learnt material that are relevant to the question set. Avoid lengthy introductions that do not directly address the question and simply provide general background material. From the outset go straight to the point of the question. To ensure continuity and relevance and to avoid drifting off the point of the question, it is important to link each major point back to the question. In this respect, planning your response at the very beginning is really important. It is recommended that you spend a few minutes planning your answer at the outset, either using bullet points that identify the key content for each paragraph, or by using 'spider' diagrams.

Managing questions

You will be issued with the examination paper for Section A and at the end of 1 hour 30 minutes that paper will collected and the Section B examination paper will be handed out. In effect it is two examination papers with a short gap. Here are some tips for good performance in the examination:
- Use a highlighter pen to emphasise the subject matter of your chosen questions.
- Use a different colour to highlight the command words.
- Always write a brief plan for each answer before you start. A plan that is not crossed out may be used by the marker in the case of an incomplete answer.
- Keep strictly to time.
- Keep to the formal essay format of sentences grouped into paragraphs arranged in a logical order. DO NOT use text-speak.
- Essays should have a brief introduction that identifies the direction of your argument and a brief conclusion that summarises your essay.
- Try to leave time to reread your essays and correct inaccuracies and spelling.
- Remember that geography at this level is about the complexity of causes, issues and problems because the interrelationships between people and environments are complex.
- You must be able to draw together information from a variety of sources. One source is not enough.
- For Section B you are expected to have undertaken research on your own away from the classroom but under the guidance of your teacher. This should be apparent in your answers.
- Remember to use examples, and to draw diagrams and maps.

Command words

It is important that you understand clearly the meaning of, and difference between, the command words or phrases used in questions. At A2 the style of questioning is more demanding than at AS and you can expect questions asking for assessment, discussion and examination. Study the list below for the meaning of all the command words.

Assess requires you to weigh up the importance of the topic. There will be a number of possible explanations and you need to give the major ones and then say which you tend to favour.

Evaluate means that, having considered the evidence and looked at the overall explanations for an issue, you must give a point of view. Credit will be given for the justification of the view that you take. It is expected that there will be more than one explanation for any issue.

To what extent and **How far do you agree** both expect explanations for and against and a justification of the view that you favour.

Discuss and **Discuss the assertion** both expect you to build up an argument about an issue and to present evidence for more than one point of view. This type of question expects you to reach a conclusion. Discussion will include both description and explanation and a summary at the end.

Examine asks you to investigate in detail, giving evidence both for and against a point of view or an opinion.

Explain requires you to give reasons or causes and show how, why and where something has occurred.

Compare asks you to point out the similarities, although many versions of such questions expect some **Contrast**. **Contrast** strictly expects the differences. **Compare and contrast** requires both similarities and differences.

Justify is asking for you to state why one opinion or explanation is better than another.

Classify expects you to group ideas or phenomena or explanatory variables into categories.

Use of diagrams and maps

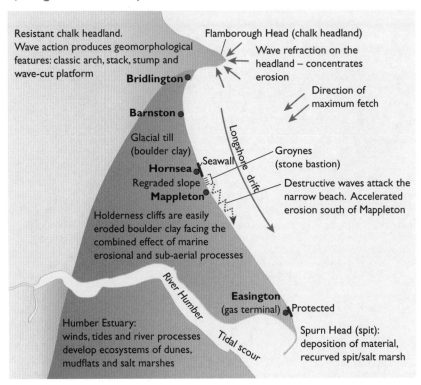

Figure 17 Holderness coast with Spurn Head

Your answers can be significantly improved by incorporating good-quality maps and diagrams. Work at including appropriate diagrams that summarise key points and

provide explanatory annotation that adds support to your text discussion. Maps and diagrams can save words, but they need to be planned as an integral part of your response rather than as an 'add-on' or afterthought at the end of an essay. In an essay that asks for an assessment of coastal management strategies, Figure 17 provides a great deal of information on the topic. It is visually effective and summarises the nature and relative location of a range of management strategies in a specific coastal environment. This would allow you time to develop your assessment of the strategies in more detail in the written part of your response.

Examiner's comments

This section contains some typical questions for Unit G3, with examples from Section A: Contemporary Themes and Section B: Individual Research Enquiry. G3A questions and the two parts of G3B should be allocated the same amount of time (45 minutes).

The **examiner's comments**, indicated by the ℮ icon, are designed to highlight the strengths and weaknesses of the candidate's answer. For each question key words are identified and advice given on how it should be approached. Examiner commentary on each student's response is broken down into annotated points using a sequence of letters and a summative statement that identifies the mark and grade awarded.

Section A: Contemporary themes

Example 1 **Extreme environments**

This question is based on Theme 1: Deserts, Key question 1.2

'The impact of human activity on the desert environment is always negative.' Discuss.   (25 marks)

ⓔ The command word 'discuss' asks you to describe and explain relevant points and build up a balanced argument with supporting detail. You should therefore give some detailed examples of ways in which desert environments are being, or have been, exploited through population growth, mineral exploitation, farming or tourism. You are likely to argue that these impacts are negative and require management, but balance your argument with examples of human activities that may have beneficial impacts, such as ecotourism or conservation.

Student's answer to example 1

The idea that the impact of human activity on desert environments is always negative is debatable. Certainly in some areas human activity can have severely negative repercussions on the fragile desert ecosystem, such as the over-cultivation and irrigation in the Aral Sea region depleting the size of the sea to a mere 10% of its original size. Yet it would be too simplistic to describe human activity as consistently negative, indeed the economic benefits arising from tourism or mineral exploitation can be viewed as positive but this is not to say that this offsets the environmental problems which frequently arise as a result of human activity.

ⓔ A well-focused introduction in which the student addresses the question directly in the first sentence and starts to build up a balanced argument — identifying human activities that are both negative and positive, but stating that on balance they are more likely to have negative impacts, especially on the environment.

Firstly a huge impact of human activity on the desert environment is that of population growth. Desert areas such as the sunbelt states of America have seen a rapid increase in population over the last few decades due to the vast untapped economic resources they proffer and the favourable climate **a**. In fact the state of Arizona in the Sonoran desert witnessed a 40% increase in population between 1990 and 2000. Yet despite increased economic benefits and lifestyle advantages achieved for those who live there, it is hard to describe the impact of population growth as anything but negative on the whole **c**. Mass population influx puts great strain on resources, particularly water, in an environment where this is already scarce. Over abstraction from Phoenix's underground aquifers resulted in a 3 metre drop of the water table; further, schemes to cope with increased demand often have severe negative repercussions **b**. The central Arizona Project siphons large quantities of water from the Colorado River for use in Phoenix and Tuscon yet this over-extraction and depletion of water supplies has led to mass ground subsidence in the Sonoran desert as a direct result of the lowering water table. Fissures up to 120m long and 5m deep have opened up in the desert, and the total area subsided is the size of North Yorkshire. The depleting availability of water due to population growth has numerous negative consequences on the desert environment. Lack of water has dried up many desert streams and the flora and fauna are being impacted due to lack of water. Indeed the desert bighorn and bobcats are no longer seen in the Sonoran desert due to the increased water stress that with population growth and the subsequent rise in demand for resources has occurred. It is therefore difficult to argue that population growth and influx has anything other than severe negative impacts upon the already fragile desert environment **c**.

e **a** Human activities and **b** their impacts (economic, social and environmental) are given in some detail and are well-located. **c** The answer maintains focus on the question.

Tourism is a booming economy in desert regions with extreme sports in Dubai and safaris in Namibia becoming increasingly popular **a**. To this extent tourism can bring significant economic benefits to desert communities through the multiplier effect whereby regeneration of a desert area for tourists has advantages for the improved lifestyles of local people **b**. The Boudin [Beouin] people in the Sahara have benefited from ecotourism to the area, not only through increased incomes and a better standard of living as a result, but also through educating tourists about their culture. To this extent tourism can provide economic and social benefits. However there is debate as to whether the economic benefits of tourism are more important than the connected environmental concerns which arise **c**. Dune and Wadi bashing in the deserts of Dubai **a** has increased erosion and land instability, enhancing the problem of desertification in the area **b**. Moreover, tourism in Morocco **a** has led to severe water shortages in the country, with the World Bank predicting a drought by 2050 **b**. Cultural drawbacks can also be witnessed as a result of mass tourism, for example the ancient monument of the Valley of Kings in Luxor, Eygpt is becoming deteriorated due to disrespectful tourists visiting the site **b**. Similarly, Native Indian tribes have been offended by the sky walk over the Grand Canyon, which they regard as disrespectful to the sacred ground of the Canyon. Evidently, the environment and social implications of tourism may undermine the economic benefits felt as this unsustainable form of tourism limits the ability of the industry to continue and prosper in the same way in the future **d**.

e **d** The candidate recognises that the impacts of a given activity can be both positive and negative and makes reference to sustainability — a synoptic point.

Agriculture in desert regions **a** creates some of the most detrimental impacts of human activity on desert environments **e**. The very nature of deserts with limited water supplies is augmented by irresponsible irrigation and over cultivation. This has occurred most obviously in the Aral Sea region of central Asia where unsustainable over-farming has led to the shrinking of the sea to $\frac{1}{10}$ of its original size, and the onset of desertification **b**. 50% of the land now suffers problems of salinity, and as well as environmental degradation. Social implications from lower yields and loss of livelihoods has led to widespread poverty in the region — over a ¼ of families are impoverished due to loss of livelihoods as a direct result of destructive farming. Agriculture practised in a negative and unsustainable way also enhances the problem of the arid environment by increasing salinity through over-cultivation, leading to the onset of further desertification. Over irrigation in Kushab, Pakistan led to 75% of the land falling victim to salinisation and expanding desertification. The consequences of human activity in the form of agriculture in desert environments is therefore frequently negative due to the unsuitability of the desert environment for farming and the unsustainable way in which it is followed. Interestingly, the most destructive forms of agriculture coincide with areas of high population density and growth as the land is abused to support people, such as in the Sahel region of Africa **f**. Therefore these areas experience not only problems posed by agriculture but also those of population increase, and so to this extent are highly destructive and negative on the desert environment.

e **e** This student appreciates that some there is variation in the degree of impacts. **f** The answer recognises that there may be a combination of human activities that accentuate the impact.

Finally, human activity in the form of mining in desert environments **a** is almost exclusively negative. Copper mining in Candeleria, Chile in the Atacama desert is highly polluting and is responsible for 10% of water pollution in the region **b**. This can be regarded as particularly destructive given the extremely scarce nature of water resources in the Atacama, with some areas not receiving rain for over 100 years. The environmental and social impacts (poor health for those who must drink the water) are heightened by the decreased sources of clean water in an already drought ridden environment. Similarly, poor rehabilitation of mines in Shoshone nation, USA has led to the release of cyanide into some water sources which was used for leaching the gold ore.

Deserts account for a $\frac{1}{3}$ of the world's mineral sources and so are rich sources of resources for developers.

Clearly, the environmental and social impacts of this are huge — the killing off of much biodiversity in the region as well as the risk faced by human population as well **b**. To this extent mining is an exclusively negative impact on desert environments as it scars the landscape, is a source of pollution and any economic benefits are only felt by TNCs and large corporations who import workers rather than local people.

Without a doubt human activity creates huge destruction to the fragile balance of desert ecosystems. Particularly environmentally, the effects of population growth, tourism, agriculture and mining are widespread and damaging. Added to this is the fact that many pressures are interlinked and occur concurrently **f** — over-farming and population growth, tourism and population growth for example, which augments and enhances the negative effects further. Whilst economic advantages are received through tourism particularly, this cannot be deemed sufficient to justify the huge environmental consequences of human activity and therefore human activity upon the desert environment can be seen as largely negative.

ⓔ **25/25 marks awarded.** This is an excellent response. It covers a wide range of detailed and well-located human activities and their impacts. Not only is it well structured, but the discussion of negative and positive impacts is maintained throughout and the student finally arrives at the conclusion that the impact of 'human activity upon the desert environment can be seen as largely negative'. The student also keeps a sharp focus on the characteristics of the extreme environment (desert) that they are discussing, as it is the fragile and special qualities of the desert environment which make them so susceptible to human activity.

Example 2 **Coastal landforms and their management**

This question is based on Theme 2(b): Coastal landforms, Key question 1.4

Examine the role of geology in the development of coastal landforms.   (25 marks)

ⓔ This question expects you to show knowledge and understanding of the effect of the character of individual rocks and lithology (hardness, mineral composition, solubility) and the effect of geological structure (bedding, dip, faulting and joints) on the development of coastal landforms. The command word 'examine' expects you to discuss the effects of geological controls in some detail, but also to look at other factors that influence the development of coastal landforms besides geology, such as differences in processes, energy levels and human intervention.

Student's answer to example 2

The coast is defined as the interface between sea and land; the area that I have studied is sediment cell 5-f, and defined by the DEFRA management plan of 1993. A number of factors work together in the production of coastal landforms, including that of geology. Geology includes both rock type and rock structure. The type of landform created involves a fine balance of processes such as sub-aerial: rainwater weathering, wind erosion; biological processes such as vegetation and animals; geomorphic — wave action, tides; human-foot erosion; and geology. The relative importance of geology depends on the balance of these processes **a**.

ⓔ **a** This is a focused introduction in which the candidate 'examines' by not only acknowledging that geological controls are important in landform development, but that other processes operate too.

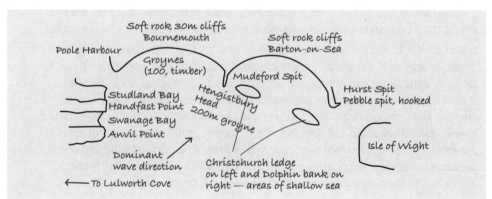

The role of geology is most important in forming erosional land forms, however it is also important in providing sediment input to create beaches, a depositional landform, as seen at Barton-on-Sea and Bournemouth. Barton-on-Sea is characterised by a horizontal bedding planes and a thick section of clay below a section of equally soft sand and shingle. Being soft cliffs allows rapid erosion by marine and sub-aerial processes. The cliffs retreat around 2–5 m a year.

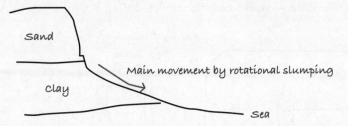

Rain water infiltrates through the sand until it reaches the impermeable clay layer (geology). There it creates an active slip zone and this small movement accounts for 89% of total land movement on the cliff; the rubble is 36% moisture, evidence of its permeability, and the cliff profile is a gentle slope. Due to the low resistance, weakening by freeze thaw and the effect of slightly acid rain erode the cliff face — evidence being an active cliff face with limited vegetation **a**. Due to limited protection from the ledge at Christchurch, wave action is high as the environment is high energy. Attrition wears away the armour formed by mass movement due to its softness.

The importance of geology is seen clearly at the series of headlands and bays that make up the eastern Purbeck coastline that lies discordant to the coast **b**.

e b The candidate uses appropriate terms effectively and **c** diagrams that clearly illustrate the effect of different lithologies on landform development.

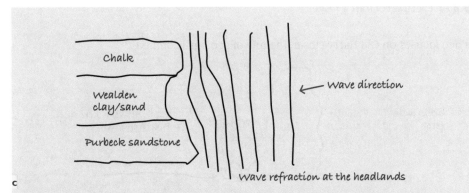

The Purbeck limestone and chalk here are more resistant than the sand and clays. Wave action, especially hydraulic action **b** which is capable of 10 000 kg force, picks out the weaknesses in the rock bedding planes and joints. The soft rock is more susceptible so rates of erosion are faster. Waves then refract **b** around the headlands, wavelength increases at the soft rock and it reduces in energy. This creates the rounded shape and forms a swash-aligned **b** beach: the difference in rock type is clearly visible as standing on the headlands one looks down on the clay/sand layer where sub-aerial processes have occurred quicker. Here rock type is most important; other factors such as wave energy are kept quite constant along this section of coast, as is the climate which contributes to sub-aerial weathering and erosion **a**.

At Handfast Point, one can clearly see the effect of rock structure in the formation of landforms. The rock type is the same — chalk, yet there has been the creation of a number of landforms — stacks and stumps included. One of these stacks is called 'Old Harry'.

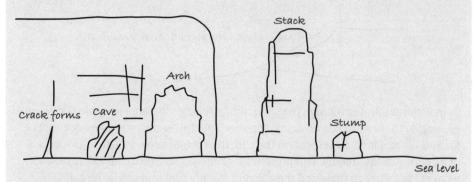

There wave action is constant, and you can see how a crack progresses into a stump. Rain water infiltrates into the weaknesses in the rock — horizontal bedding planes and vertical joints. Continued sub-aerial processes and formation of a wave cut notch by hydraulic action and corrasion lead to rock fall as it is not supported below. Rock falls off in blocks, evidence of the importance of rock structure. Rock type is important to a degree here as chalk allows some resistance to the wave action. For example, these structures cannot be seen at Barton-on-Sea, where the rock type is much softer **d**.

🅔 **d** An effective contrast is made.

If one focuses on Old Harry closer, the role of geology is limited.

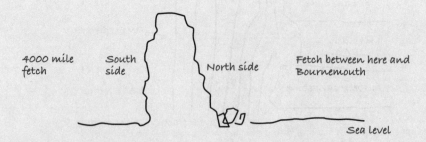

Even though geology is the same, the aspect is affected by fetch **b** as a bigger fetch creates stronger waves and more rapid undercutting. The south side is white, steep and undercut and the north side is less steep, has vegetation growth and the armour not yet removed. This shows even geology is limited in its importance in forming landforms **a**.

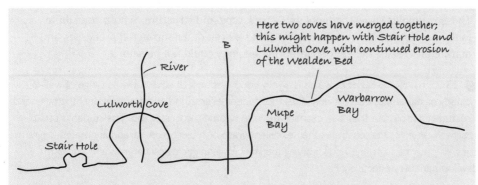

On the concordant Jurassic coastline, geology allows for the creation of coves and bays. 10 million years ago Europe and Africa collided, causing rock here to fold — known as the 'Lulworth crumple'. The bands of Purbeck and Portland limestones **e** are 'S'-shape folded, meaning there is only a thin layer of limestone through which the waves must breach **f**. During the Flandrian transgression **b** 12,000 years ago ice melted and a river of meltwater flowed at a point indicated on the map. This cut a weakness into the rock where wave action could attack easier.

ⓔ **e** Actual rock types are correctly identified as well as **f** structural controls.

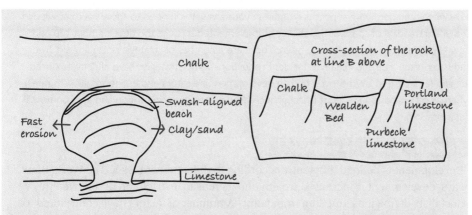

Once the wall was breached there was rapid erosion of the less-resistant clay and sand. The extent of the cove is limited by resistant chalk behind. Sea wave action is also important as wave refraction in the cove means wave energy is lost, and it deposits sediment, forming a swash-aligned beach in the cove. The early stages of this development can be seen at Stair Hole.

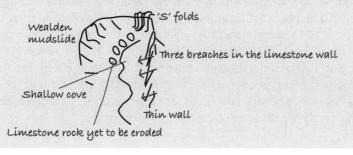

These examples show that geology — rock type and structure — both contribute to the development of landforms, but that without other sub-aerial processes and marine processes, the effect of differing geology could not be seen **a**.

(e) **25/25 marks awarded. a** This is a very good response. It covers a wide range of well-located coastal landforms. Not only does the student refer to the effect of different lithologies and structural controls, but they also examine the role of other processes too. The student's use of geographical terms demonstrates their excellent grasp of subject material and annotated diagrams are well integrated into the text, showing the link between geological controls and landform development very effectively.

Example 3 **Development**

This question is based on Theme 4: Development, Key question 1.6

'Strategies for reducing the development gap are not always effective.' Discuss.　　　(25 marks)

(e) The command word 'discuss' asks you to describe and explain relevant points and build up a balanced argument with supporting detail. In your answer you need to show your knowledge and understanding of some of the strategies implemented to close the development gap and how effective these strategies have been. There are a number of strategies including aid, free and fairer trade, FDI and initiatives for debt reduction. Some strategies have effectively closed the gap (as indicated by improved human development indicators and economic growth rates) whereas others have widened it (through increased inequalities, corruption and environmental deterioration).

Student's answer to example 3

Development is defined differently according to different sources **a**. The World Bank says 'development is increased wealth and increased distribution of that wealth'. Socialists define it as 'freedom from want'. A number of factors have contributed to the development gap — which is a development differential between, and within, countries **b**. For example, the USA GDP is four times higher than the international average, and it used to be only two times the national average **c**. 85% of the wealth in Sierra Leone is held by 1% of the population. This development gap may correlate with social development as Sierra Leone has an HDI (Human Development Index) rank of 173 in the world. An appreciation of the value of not having such a great gap has led to a number of strategies to reduce it. These include debt relief, aid, free trade, fairtrade, FDI and remittances **e**. A measure of their effectiveness may be the success of the Millennium Development Goals (MDG) of 2000, in which 189 countries are taking part. I agree some strategies have not helped development and thus reduce the development gap. Some have even hindered development **e**. The extent of their success varies between countries, as each country has a different history, climate and supply of natural resources such as oil or rubber.

🄔 This is an effective introduction in which the student has defined the terms **a** development and **b** development gap and **c** provided some examples. They start to build up a balanced argument, **d** identifying a range of strategies, while acknowledging that not all of these have helped to reduce the development gap and suggesting that **e** 'some have even hindered development'.

The IMF decided to help Heavily Indebted Countries (HDC) by rescheduling their debt **d**. In order to do this, the country had to agree to certain Structural Adjustment Policies (SAPs). In 1986, Senegal began their SAP. As an example they were told to grow more cash crops, like coffee and tobacco, in place of staple crops. The theory is these crops can be sold, earning export dollars, and this money could then be invested in development such as infrastructure and healthcare. Unfortunately, many countries were told the same thing so the value of their crops on the world market decreased and farmers did not see the return on their investment. Land had to be sold, which increased rural poverty, increased urbanisation and reduced the amount of staple food produced **e**. The country then had to import food from elsewhere, so there was less money to invest. 10 years later, child malnourishment increased to 22%. Malnourished children means low productivity at school and lower potential for employment and therefore good health in the future. Some Non Governmental Organisations (NGOs) manage debt relief more effectively **f**. 'Debt for Nature' means the debt is taken from the IMF. The NGO accepts domestic currency and the money they make is used to make them not deforest. In the Stern report he said the benefit of not deforesting is valued at $5 billion a year. Things like this help the country come out of debt and also improve their environment — needed for sustainable development.

🄔 The student balances their discussion by identifying **e** strategies that have increased the development gap with **f** those that have caused it to narrow.

Many say 'Trade not Aid'. Trade may be more effective at increasing development as it creates economic activity. However some aid projects are in place. Some are successful and some less so. After Live Aid the BBC found 80% of the aid in Somalia went missing **e**. Corruption there leads to stealing of food aid. It can be sold cheaply or for free — undercutting local farmers and leading to the same problems as with cash crops. The UK gave Samoa a multi-million pound hospital. This was inappropriate aid as the infrastructure was not yet in place to allow access to it; nor did the government have enough money to afford equipment or staff.

Water Aid helps development and is successful **f**. Covered wells in Zambia have helped 5000 people. They work by hand pumps. The village is not reliant on diesel or electricity or expensive replacement parts. Being covered, these wells are not contaminated with human faeces, chemical fertilisers or dead animals. Prevalence of water-borne diseases decreased like diarrhoea which made parents less productive and killed babies and children. Water Aid also built compost toilets in northern India, which convert human waste into clean safe manure for crops. A wider variety of crops could then be grown — either to sell or for improved nutrition.

The World Trade Organization (WTO) is dogmatic in its approach to free trade. Free trade is created no matter what the circumstances. In the case of the Windward Island bananas, free trade hindered development **e**. A quota had to be lifted on

non-Windward Island bananas into the EU (who consume 2/3 of all bananas). This quota was in place to protect the islanders' trade, as 51% of exports were bananas, and it created 36% of all the jobs. Without the quota, the American plantations could access the market. Companies like Del Monte have economies of scale so their bananas are sold cheaper than Windward Island bananas. As a result, consumers buy USA bananas and now half of the farmers in the Windward Islands are on less than $1 a day. Conversely if the $600 per hectare subsidy on USA cotton was removed, USA cotton prices would increase 25% on the world market and West African farmers' income would be $190 million f. With the subsidy, their cotton is more expensive to buy, despite it being produced at 26p a kilo compared with 96p a kilo in USA.

The alternative to free trade is fair trade. Fair trade produce is a social movement with a market-based approach f. It is worth $5bn a year and its value increased 22% a year (in 2007 before the recession).The farmers receive 2–18% of the value of the product and the income does not fluctuate according to the value of the markets. offering them a stable, reliable source of money. The 'fair trade premium' goes to the organisations who manage the farmers, and they invest in things to help developments, for example buying $200,000 of chemical fertilisers in view of rising oil prices — so that the farmers' costs would not increase in the near future. They may also invest in pension funds, this ensures children can afford education even when the parents can no longer work, and this improves educational attainment and potential for employment and escape from the poverty trap. The increased income may be used to invest in a tractor or processing plant to increase productivity or add value to the primary good. Unfortunately, trade blocs prevent the farmers from selling secondary goods e. Raw coffee beans from Africa into the EU are taxed at 9%. Roasted coffee beans are taxed at 16%. Roasted beans within the EU are cheaper for consumers so African roasters do not benefit. Progressing to secondary industries would improve development as it creates cumulative causation and the multiplier from the extra employment created and forward and backward linkage industries like a bottle plant or label designer. Even with strict quotas and tariffs on goods, I still believe trade can help a country develop. China set up special economic zones (SEZ) which had reduced taxes, so encouraged foreign direct investment (FDI). This is a form of globalisation, and the taxes and employment generated help contribute to the 13% rise in economy of China per year. Malaysia controlled their FDI so only primary products and products that could not be manufactured domestically could be imported (through quotas and tariffs on other products). Domestic industry then could profit, innovate and improve once the comparative advantage was developed — cheap or willingness of labour or economy of scales; they globalised and could dominate their market. A similar thing happened in South Korea with the company LG which eventually globalised to South Wales. The success in Malaysia can also be attributed to political will. (The dictator Mohamed could better enforce policies; he was more of a benign dictator so changes were made for the greater good of the country. Unfortunately, the same is not true for all countries.)

In Equatorial Guinea, a 'city' has just been completed, complete with heated pool, heliports, artificial beaches and an 18-hole golf course e for the AV summit in the last week of June. Spending like this is known as a 'white elephant' and is detrimental to the local population — here 75% of the population are on $1 a day. The government

chool children away from school early as it feared they would cause
Decisions like this can seriously hamper any previous progression in

vered all areas of development from social health, gender equality,
l sustainability and economic development. A combined effect of
is seen in the success so far. The number of people living on less
has decreased from 1.4 billon to 833 million — a success — and there
ntil the deadline in 2015. Unfortunately, 40% of low-income countries
t their MDGs on infant mortality rates and no low-income country will
vironmental sustainability goals. Currently only 0.2% of total GDPs
untries contribute to the MDGs, less than the 0.7% target. Success is
nited. This is just an average, however, and depending on how the policy
implemented in many cases development has improved. The most successful
measures seem to be the bottom-up, small-scale, participatory ones, which help to
improve development naturally and slowly, with economic and social sustainability
g. Therefore I agree that strategies for reducing the development gap are indeed
not always effective, unfortunately. The reasons for their lack of success vary
in countries, whether it be lack of democracy, corruption or unsustainable debt
(classified as 200% of total export income).

e **25/25 marks awarded.** This is another excellent response. The student has written a very detailed and wide-ranging answer, with a balanced discussion throughout and **g** has been able to arrive at an overall conclusion about the effectiveness of a range of strategies.

Example 4 **Emerging Asia**

This question is based on Theme 6(a): China, Key question 1.4

Describe and account for the increasing inequalities between rural and urban areas in China. (25 marks)

e This question contains two command words, 'describe' (identify distinctive features and give descriptive, factual detail) and 'explain' (give reasons or causes) and it is essential that you respond to both of these. You should give details of, and reasons for, the increasing inequalities between rural and urban areas in China.

Student's answer to example 4

China experiences severe inequalities between rural and urban areas largely as a result of migration and urbanisation and its recent and rapid economic development. Urban to rural incomes follow a ratio of 4:1, highlighting the economic inequalities and disparities between urban and rural regions **a**. Rural regions experience greater social and economic inequality compared with urban areas and often have a much lower standard of living. Despite 200 million people being lifted out of poverty in the last three decades, and greater economic liberalisation, a static social policy in China has enhanced rural to urban inequalities further.

e This student has provided some detail in their **a** description and **b** explanation.

Over 80% of migration within China is to urban areas and less than 18% is to rural areas. As a result rural China faces problems of a changing social structure as over 70% of migrants to urban areas are of working age **b**. In light of this urban China has developed rapidly and the standard of living for urban residents has increased likewise. In contrast the economy of rural areas has declined due to the loss of able-bodied young workers who are required for the labour-intensive agricultural work of rural regions. Incomes for rural families have declined as a result in contrast to the climbing income of urban workers. Indeed the income of urban Shanghai was 12 times that of rural Guizhou in 2007, highlighting the increased economic inequalities which have resulted from mass migration to urban areas as a result of increasing urbanisation **a**. Rural areas also face social inequality as a result of migration; not only are there between 13 and 20 million children in rural China where at least one parent has moved to urban areas for work indefinitely, but rural migrants in urban areas are discriminated against due to the Hukou registration system which prevents access to basic welfare and health and education for migrant workers. Rural areas (and citizens) therefore face increasing social and economic inequality in comparison with urban areas, which are developing rapidly and becoming more prosperous.

Economic policies of the government in China are largely accountable for much of rural–urban inequalities **b**. The establishment of Special Economic Zones such as in Shenzen and of 'Open Cities' like Shanghai in the 1980s benefited urban areas economically, at the expense of rural regions. It must also be noted that these benefits were not felt by all urban areas, the economic benefits were felt by those cities in the more populous southeast and the coast as opposed to the country as a whole. As a result there is also spatial inequality in China as well as rural–urban inequalities **c**. As a consequence of the rapid industrialisation ushered in by these government initiatives, urban areas on the east coast benefited disproportionately in comparison with rural and inland areas Shanghai became the richest province and by 2007 was receiving $40 billion in foreign direct investment. This therefore accounts for the economic superiority of urban areas in comparison with the relative poverty of rural regions.

e c The student identifies that there are spatial variations between east and west, with rural areas to the west being more disadvantaged.

Urban expansion has played a significant role in enhancing regional inequalities **b**. 47% of China's population is now urbanised and the encroachment of urban areas onto agricultural and rural land — to the extent of 33% per capita loss of rural land since 1990 — has detrimentally affected rural economies which are highly dependent on agriculture. The urban area of Shanghai has expanded by 486% and has severely reduced agricultural land in a country which has 20% of the world's population yet just 7% of farmland. Whilst urban areas, as of yet, are not experiencing the downsides of urban expansion on the food security of China and merely benefit from the economic advantages of urbanisation, rural regions are suffering. 70% of rural economies in China rely upon agriculture, and particularly in the arid north where land is less productive, reduced yields are impacting on rural standards of living.

Further, although urban expansion offers the potential for a wider market for rural areas to trade with, lack of developed infrastructure hinders their ability to do this and so inequalities between rural and urban regions remain entrenched. Government population policies have further enhanced inequalities in China **b**. The One Child Policy of 1978, whilst not enforced as strictly in rural regions, has severely depleted the workforce of rural China. Whilst urban residents are affected to a lesser extent, fewer hands on the farms has decreased agricultural productivity and reduced rural living standards further. Moreover, the problem of an ageing population — 16% will be over 60 by 2020 and the dependency ratio will have fallen to just 3 workers to 1 person to support — has been felt by rural areas more significantly. In rural Sechuan the dependency 4-2-1 pattern which a single child must support has a prevalence of 30% and is expected to rise. Social hardships of rural families (where the grandparents traditionally live with the children) results in a lack of social security for an ageing population. Poor rural access to healthcare — only 3.8% have access — enhances these problems further.

Additionally, globalisation of China has benefited urban areas disproportionately to rural regions **b**. Whilst urban incomes have increased six-fold due to increased FDI (at $93.7 billion in 2008) in the 'Open Cities' of the east coast and have created a larger middle class and more prosperous society, rural incomes have remained static and even declined by 12% in the impoverished region of inner Mongolia. Urban areas such as Shanghai have benefited from an expanded global market, and residents of the city have experienced a 60% increase in standard of living since 1990.

Evidently, despite the vast economic growth which China has undergone, with its GDP growing to a staggering $3.5 trillion dollars by 2009, the effects of this have been disproportionately spread. Whilst urban areas have witnessed increased standards of living as a result of China's development, rural regions have suffered greater inequality as a result of rural to urban migration, an ageing population and declining agriculture. Chinese development has been atypical of developing countries in many respects, but perhaps significantly for the disparate and inequal share of benefits of its development between urban and rural areas.

(e) **25/25 marks awarded.** An excellent answer. This student has described urban–rural inequalities in detail, often with statistical support, and offered a range of linked reasons for these.

Section B: Individual research enquiry

Example 5 **Microclimates: valley microclimates**

(a) Identify and justify methods of presenting data that could be used in an investigation into valley microclimates.
                                                                                                    (10 marks)

This question contains two command words that assess your ability to apply understanding and evaluation (AO2) of a range of skills and techniques (AO3) used in presenting data for your chosen topic. The commands are to *identify* (point out and name from a number of possibilities) and *justify* (explain why your choice is better than the possible options) and it is essential that you respond to both command words. The phrase 'could be used' gives you the opportunity to mention other suitable methods of presenting data that were available to you, even though you chose not to use them in your own research.

Student's answer to example 5

In my investigation into valley microclimates I presented my data in many ways. I used a polar graph to present temperature and aspect **a**. This was effective as it gave me a 360° overview of my data. However, the segments got proportionally smaller the nearer to the middle of the circle. Therefore if data were cramped together in the middle it made it difficult to read **b**. I used a dispersion graph to show temperature differences on either side of the valley **a**. This allowed me to compare similar sets of data. In addition, it allowed me to analyse upper and lower quartiles and the median. The median moreover was both visual and important to the analysis side of my investigation.

[Student's dispersion diagram went here]

However, as this diagram shows, the dispersion graph would be ineffective when trying to present data such as aspect — where a 360° overview is needed **b**.

I used a scatter graph to present wind speed and altitude as it is continuous data **a**. The scatter graph allowed me to compare correlations, see trends and made data clearer. However, if the data were cramped together it made the analysis of the readings more difficult **b**. From it, however, I could find Spearman's rank — which I used Excel for. This mathematical equation gave me a statistical link and allowed me to put a value on the strength of the correlation. Spearman's rank is an accurate way of assessing as its accuracy depends on the number of data sets — not just the value of the correlation by eye **c**.

I used a proportional arrow **a** to present wind direction and wind speed. I used tracing paper — and placed it over a map of my valleys — Cwm Idwal and Nant Ffrancon. This was a visual way of presenting my data and made my data clearer **b**.

e **8/10 marks awarded. a** This is a strong response, identifying four methods of presenting data that are linked effectively to the topic with **b** a justification of each one. **c** The reference to the technique Spearman's rank is not appropriate here as this is a method of analysis, not a method of presenting data. This is a Level 3 answer and would gain a grade A.

Example 6 **Rivers: river sediments**

(a) Outline how information may be collected in an investigation into river sediments. (10 marks)

e This question contains the command word *outline* (give a brief summary of the main characteristics), which assesses your ability to apply understanding and evaluation (AO2) of a range of skills and techniques (AO3) used in collecting information for your chosen topic. You should give a brief summary of how information or data could be obtained (primary and/or secondary data collection) in the context of your selected research on river sediments. The phrase 'may be' gives you the opportunity to mention other suitable methods of collecting information that were available to you, even though you chose not to use them in your own research.

Student's answer to example 6

In an investigation of river sediments, random sampling can be used to choose the sites to be investigated along the river **a**. It is necessary to investigate around ten sites to gather the information needed for the investigation.

Once the sites have been chosen you need to collect the data. To collect these you must measure the width, depth, velocity, bed load and sediment load **b**. To measure the width you measure from each bank side. To get the full width, you need to measure from the top of each bank, where the water would be after heavy rainfall. To measure the depth, you need a metre rule and a tape measure at 0.5 m. Then at every 0.5 m you measure the depth with a metre rule recording each time until you get to the other side.

[Student's diagram went here] **c**

To measure the velocity, it can be either done with a float and measuring out a 10 metre stretch along the river and timing how long it takes for the float to go the 10 metres and repeating three times **b**. Or you can use a flow meter, which is placed in the river, in a central location to obtain an average reading. The flow meter allows for a direct reading by recording the numbers of counts over 1 minute from which the river's speed can be calculated by using a conversion chart. The readings were repeated three times.

To measure the bed load **b**, you use a random area sampling and collect 20 samples of the bed load **a**. Then for each measure the y-axis of the rock and also the Cailleux roundness index. To do this you choose the pointiest end and measure it with the Cailleux index. Also you can categorise the shape with the Powers shape index. With this you can tell whether the rock is angular or rounded.

To measure the sediment load you need a plastic bottle with two tubes: one out of the river and one facing the flow **b**. You hold the bottle in the middle of the river between the bed and the top and keep it there until the bottle has filled up. Then label the bottle.

ⓔ **6/10 marks awarded. a** This student has mentioned the sampling techniques used, but not elaborated on their choice. **b** The answer is competent but rather generalised, lacking development of the methods used and specific detail, particularly with respect to the aim of the investigation. **c** The inclusion of a diagram often aids the explanation. This is a Level 2 answer and would gain a grade C.

Example 7 **Microclimates: valley microclimates**

(b) Outline the main findings of your personal research into valley microclimates and evaluate your methods of obtaining information.   (15 marks)

ⓔ As stated in the A-level specification, question (b) will examine the findings of the candidate's own research, so it is essential that you can recall your findings in some detail and provide supporting evidence and statistics. This question also asks candidates to evaluate the methods used to obtain information. It is essential that you respond to both command words as an unbalanced response will restrict your mark to Level 2 (5–8 marks) or lower.

Student's answer to example 7

My investigation into valley microclimates endeavoured to discover whether south-facing slopes had higher ground temperatures, wind speed was higher at the top of the valley and if winds acted in an anabatic **a** fashion **b**.

My first hypothesis of 'south facing slopes had higher ground temperatures' used 37 data sets **c**. From my background research I knew that during the day the south facing slopes of Cwm Idwal had a lot of sun due to the low surface area. At night this resulted in a temperature inversion (cold air on the valley slopes and warmer just above) this creates a visible fog line along the water condensation line **d**. Therefore I used a sky watch with a built-in anemometer and thermometer to measure the temperature. According to my systematic sampling technique I stopped every 200 m and held the sky watch 2 cm off the ground. To make it fair I allowed 1 minute to adjust to my surroundings and measured it in the shade so the elastic didn't get hot. I did these methods every time to ensure accuracy. Moreover, data were taken simultaneously, by a set of geographers taking separate routes around the valley to ensure maximum precision **e**.

I presented my temperature data in a dispersion graph and a polar graph (aspect was taken using a compass). The median of the southern slopes was 23.7 °C compared with 20.3 °C of my northeast slope **d**. This shows how effective my background research was and it was the most successful conclusion of my three hypotheses. My methods were valid and fair and therefore my hypothesis could be proven correct **e**.

My second hypothesis of 'wind speed was higher at the top of the valley' took into account 38 data sets **c**. I used a sky watch to measure my wind speed by holding it at arms length. I did this three times and took the average. By doing the same procedures each time it ensured validity and accuracy in my methods of obtaining the information **e**. According to my background information high wind speed areas usually accelerate during the day at a maximum in mid afternoon — which is when I measured my results. This ensured I was taking the data at the opportune time. I used a scatter graph and Spearman's rank to present my findings. Spearman's rank showed a reading of 0.245 compared with 0.274 for a 95% chance of likelihood **d**. This suggested that my anomalies — six readings of zero — marred my results. If I was to do the investigation again I would use two or three anemometers to eradicate anomalies. My methods of obtaining were seemingly correct however, repeating on a different day would fully ensure accuracy. However, my data —although not strongly correlating — did suggest wind speed was higher at the top of the valley.

My final hypothesis was 'winds will act in an anabatic fashion' **c**. From my background research I knew that the sun warmed valley slopes in the day and that air rose and created an upward-sloping wind. To measure this I blew bubbles and put the anemometer above my head to see the direction. This worked well — but a sudden gust of wind could mar my results and there is a high risk for human error.

However, I used the same procedures each time to ensure accuracy **e**. I used a proportional and directional overlay on a map of Cwm ldwal and Nant Ffrancon. For the data on the eastern side — it showed what I expected — a north/northeast/east wind which met my hypothesis — and seven sets of data correlated this **d**. For my data on the western side of the valley, it showed the wind funnelling along the valley, which met my hypothesis to some extent **d**. Unless weather is almost completely anticyclonic general wind will override anabatics.

Overall, in order to truly say my hypotheses were correct I would repeat my investigation on a different day — in different weather conditions. Perhaps I would do it in the winter and on a non-anticyclone day to ensure opposite conditions from the ones I had to begin with. I did all I could for validity and accuracy of my data and methods of obtaining data — by using the same procedures each time, making sure the data were tested before hand — and any equipment was calibrated. The biggest flaw in my investigation was the weather conditions. I knew that microclimates only existed for a short period therefore I ensured data were collected simultaneously so that results obtained would be accurate **e**. Therefore, I believe my findings were successful and fair.

e **15/15 marks awarded**. This is a very competent response and the student has written a considerable amount in the time available. **a** The answer uses geographical terminology appropriate to the topic. **b** At the outset the student has clearly stated the aim of the research enquiry. **c** Three separate hypotheses are stated. From the detail provided in this response it is clear that the student has undertaken the necessary research and understands it. **d** The answer has also responded directly to the key command words in the question as findings are outlined in considerable detail and are often quantified. **e** Methodology is detailed and sophisticated and clearly evaluated including reference to ways in which it could be improved. This is a Level 4 answer and would gain a grade A.

Example 8 Leisure and recreation: leisure, recreation and urban regeneration

(b) Summarise the main conclusions of your personal research into leisure, recreation and urban regeneration and discuss how these conclusions support your initial aims. State the title of your investigation.   (15 marks)

e As stated in the A-level specification, question (b) will examine the findings (including the conclusions) of your research, so it is essential that you can summarise your conclusions in some detail and provide supporting evidence and statistics. This question also asks you to discuss how these conclusions support your initial aims. It is essential that you respond to both command words, as an unbalanced response will restrict your mark to Level 2 (5–8 marks) or lower.

Student's answer to example 8

'As You Get Closer to X, Urban Regeneration, Leisure and Recreation are More Apparent.'

My personal research followed the hypothesis above, I was investigating the link between leisure and regeneration in relation to urban regeneration, after collecting all of the information in question (a) I looked at me **e** results and concluded that the hypothesis was very accurate.

My pedestrian count showed that the number of people in the regenerated area was considerably higher than in the non regenerated area **a**, this linked directly to the fact that regeneration brought with it facilities such as canoeing, sailing and many more along with cafes, restaurants and shops, meaning the reason for the difference in the amount of people is largely due to these shops and facilities.

e **a** The student concludes that there were more people in the regenerated area, but does not provide any figures to support this finding.

My litter count showed that there was less litter in the regenerated area **b** although there were more people, meaning the council were **e** taking good care of maintaining this area regularly. In the other areas there seemed to be far more litter however very few people about, being almost overshadowed by the regenerated area.

e **b** Again a conclusion is drawn but the student does not provide any figures.

The car registrations in the regenerated area showed that the majority of cars were new and expensive whereas in the other areas they were quite old and used cars, this supports that **e** the more affluent people owning newer, more expensive cars have more disposable income, allowing them to participate in leisure and recreation activities which is why they were in the regenerated area of X. **c**

e **c** Conclusions are also generalised.

Census data showed that the people living in the regenerated area had high paying jobs or a dual income with a partner suggesting that accommodation is expensive. **c**

The land use maps were very useful because they showed the scale of regeneration, the amount of regeneration that had been done and the number of leisure and recreation facilities. I could also plot my pedestrian count on the map showing what facilities people were close to. **c**

Questionnaires suggested the majority of people were active and were using the facilities on the river. **c**

My hypothesis proved to be very accurate **d** and followed the trend of activities at X.

e **7/15 marks awarded.** This essay outlines the investigative techniques used and the conclusions reached, but these are all generalised. Because there is no effective evaluation of those conclusions as required by the question, the student cannot access the higher marks because they have failed to respond to both command words, restricting the mark to Level 2 (5–8 marks) or lower. **d** The last sentence provides a token evaluation. **e** Also, the quality of the language, particularly the punctuation, is variable and detracts from the flow of the essay. Some of the results could have been critically examined to test their veracity. This is therefore a Level 2 answer and would gain a grade D.

Knowledge check answers

1 Strategies include planting trees for shade and to act as windbreaks; controlling grazing; re-seeding areas with new varieties of drought-resistant crops; using 'magic stones' to collect moisture.

2 Plants have short roots so as to avoid the permafrost and small leaves so as to limit transpiration. Plants are also dark and hairy. The darkness of their flesh absorbs solar heat, and the hair helps to trap the heat and keep it close to the surface of the plant. Some plants even have dish-like flowers that track the sun. Animal adaptations include short and stocky arms and legs, thick, insulating cover of feathers or fur which changes colour — brown in summer and white in winter, thick fat layer gained quickly during spring in order to have continual energy and warmth during winter months, and many tundra animals have adapted especially to prevent their bodily fluids from freezing solid.

3 Plucking is concentrated on the backwall. The material eroded through plucking (moraine) is then used to abrade the base of the cirque, deepening it further.

4 Glacial deposits are made up of sand and clay mixed with larger clasts and boulders, there is no sorting of the deposits and the clasts are generally angular. Fluvioglacial deposits are more rounded, as they have been subjected to processes of attrition and abrasion, and are sorted because of the changing competency and capacity of meltwater streams.

5 Dynamic equilibrium refers to the balanced state of the coastal system when inputs and outputs are equal. If one element in the system changes because of some outside influence, this disturbs the equilibrium and affects other components. This is known as feedback, which can be either positive or negative.

6 Hard engineering uses structures to control coastal processes such as seawalls, gabions, rip-rap and groynes, which are expensive and often unsightly. Soft engineering includes beach replenishment or managed retreat, strategies that work with nature and are more sustainable.

7 The main global location of low pressure is at the equator and the main locations of high pressure are 30° north and south of the equator. These positions change seasonally with the changing position of the overhead sun.

8 A distinctive dry season.

9 Convection rainfall is rain that occurs due to the heating of the ground causing warm air to rise, cool and condense. The two other types are frontal and relief (orographic) rainfall.

10 MEDC, NIC, OPEC, RIC, CPE/FCC, LEDC, LDC

11 The difference in rankings reflects that some countries are more 'developed' in some aspects of development than in others. Saudi Arabia has a much higher GNI ranking compared with its HDI and GDI scores. This suggests that the wealth generated from oil revenues (giving the high GNI per capita ranking) is not being spent so much on health and education compared with other countries and that the gender gap is particularly wide.

12 FDI is a direct investment that occurs across national boundaries when a TNC or global company sets up a branch or invests in a firm in another country.

13 Organization of the Petroleum Exporting Countries. Members include Algeria, Angola, Ecuador, Iran, Iraq, Kuwait, Libya, Nigeria, Qatar, Saudi Arabia, United Arab Emirates and Venezuela.

14 There are 32 post-completion point countries, including Bolivia, Ethiopia, The Gambia, Ghana, Haiti, Malawi, Mali, Niger, Uganda and Zambia.

15 NAFTA (North American Free Trade Agreement); ASEAN (Association of Southeast Asian Nations). The EU/NAFTA/ASEAN triad dominates the world economy.

16 Sub-Saharan Africa

17 The proportion of hungry people has declined from 32% in early 1990s to 28%. By 2007 73% of children were immunised against measles, but 1 in 7 still die before their 5th birthday. The prevalence of HIV has levelled off. Sanitation coverage has improved.

18 South Korea, Taiwan, Hong Kong and Singapore

19 Satellite and fibre optic communication have reduced the cost of communication and led to the growth of the internet and mobile phone.

20 Containerisation has led to cheap and more efficient modes of transport. Cheap air travel using jet aircraft has transformed the transport of high-value, low volume goods.

21 Economic reforms from 1978, particularly Deng Xiaoping's 'open door' policy which encouraged TNCs to locate manufacturing branches in China.

22 Labour costs are lower. India has a large English-speaking workforce. There are ICT skills shortages in some developed countries.

23 Glocalisation refers to the sourcing of materials locally and adapting global brands to local conditions.

24 The South-to-North Water Diversion Project in China is the largest of its kind ever undertaken. The project involves drawing water from southern rivers and supplying it to the dry north. Planned for completion in 2050, it will eventually divert 44.8 billion cubic metres of water annually to the population centres of the drier north such as Beijing. The project will link China's four main rivers — the Yangtze, Yellow River, Huaihe and Haihe.

25 The SEZs were deliberately located far from the centre of political power in Beijing, minimising political influences. More specifically, the original four zones were sited in coastal areas of Guangdong and Fujian, which had a long history of contact with the outside world through outmigration, and at the same time were near Hong Kong, Macao and Taiwan. The choice of Shenzhen was especially strategic because it is situated near Hong Kong, the key area from which to learn capitalist modes of economic growth.

26 Since 1980, the PRC has established special economic zones in Shenzhen, Zhuhai and Shantou in Guangdong Province,

Xiamen in Fujian province and designated the entire province of Hainan a special economic zone.

27 Examples include Chery Automobile, Dongfeng Motors (car manufacturers), Pacific Century Cyberworks (telecommunications) and CNOOC (oil company).

28 The children of young adults are brought up by their grandparents. Rural populations are ageing which impacts on agricultural productivity. The geographical separation of many young adults from their ageing parents has posed significant challenges to traditional patterns of familial support to rural older people. The social insurance system has not been set up in most of the rural areas, and the new collaborative medical care system is now at the pilot stage, hence the old age security and the medical care of the farmers cannot be guaranteed in rural areas, especially in western regions.

29 As of 2009, 40 ecocities were in development including Dongtan and Huangbaiyu.

30 Costs include: skilled jobs are filled by imported Chinese labour; Chinese aid to African countries is tied aid; factories are Chinese-owned and prevent indigenous firms from developing and the development of a manufacturing base (the export of raw materials from African countries still dominates trading patterns). Benefits include: aid, employment opportunities, increased capital, technology and expertise.

31 In addition to environmental concerns (coal is one of the dirtiest hydrocarbon fuels), coal cannot meet all of India's energy needs. The transportation industry requires oil, and much of India's coal is not of the type needed in steel and other industries.

32 There are insufficient roads, bridges and airports due to under investment and political inertia. Congestion and poor roads cause economic losses estimated at US$6 billion per year.

33 Rapid economic growth has been due to the expansion of the service sector rather than to the growth of manufacturing. Business links to North America and Europe dominate rather than to Japan and other Asian nations.

34 Examples include Aditya Birla Group, Hindustan Computers Ltd, Infosys Ltd, Tata Group, Videocon Industries and Wipro Tech.

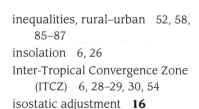